PROTOPLASMATOLOGIA
HANDBUCH
DER PROTOPLASMAFORSCHUNG

HERAUSGEGEBEN VON

L. V. HEILBRUNN UND F. WEBER
PHILADELPHIA   GRAZ

MITHERAUSGEBER

W. H. ARISZ-GRONINGEN · H. BAUER-WILHELMSHAVEN · J. BRACHET-BRUXELLES · H. G. CALLAN-ST. ANDREWS · R. COLLANDER-HELSINKI · K. DAN-TOKYO · E. FAURÉ-FREMIET-PARIS · A. FREY-WYSSLING-ZÜRICH · L. GEITLER-WIEN · K. HÖFLER-WIEN · M. H. JACOBS-PHILADELPHIA · D. MAZIA-BERKELEY · A. MONROY-PALERMO · J. RUNNSTRÖM-STOCKHOLM · W. J. SCHMIDT - GIESSEN · S. STRUGGER - MÜNSTER

BAND I

GRUNDLAGEN

2

BIOCOLLOIDS AND THEIR INTERACTIONS

WIEN
SPRINGER-VERLAG
1956

BIOCOLLOIDS
AND THEIR INTERACTIONS

BY

H. L. BOOIJ AND H. G. BUNGENBERG DE JONG
LEIDEN   LEIDEN

WITH 159 FIGURES

WIEN
SPRINGER-VERLAG
1956

ISBN-13: 978-3-211-80421-6   e-ISBN-13: 978-3-7091-5456-4
DOI: 10.1007/978-3-7091-5456-4

Biocolloids and Their Interactions

with Special Reference to Coacervates and Related
Systems

By

H. L. Booij and H. G. Bungenberg de Jong

Department of Medical Chemistry, University of Leiden

With 159 Figures

Contents

1. The Rôle of Colloids in Biological Systems

An analysis of "living matter" reveals that protoplasm is a very complicated mixture of organic and inorganic substances. Proteins, nucleic acids and lipids are the normal components of this mixture. Many products of protoplasm (e. g. cellulose) fall in the class of the polysaccharides. As these substances frequently show the properties of colloidal systems our first task will be to give a very short survey of colloid science, which will enable us to classify the various biocolloids.

Two conflicting lines of thought have governed colloid chemistry during its history. These ideas are connected with the two causes which might explain the anomalous behaviour of colloids (abnormally low osmotic pressure, very low diffusion velocity, etc.):

a) the dissolved substance is in true solution, but it has a very high molecular weight,

b) the substance is present in a very fine, but still polymolecular subdivision (thus forming a "pseudo-solution").

Physical chemists intensively studied colloidal solutions of gold, As_2S_3 and other substances. Their successes caused an overestimation of the second line of thought and colloid chemistry became the physical chemistry of two-phase systems; one of the phases being dispersed to colloidal dimensions in the other phase. Here the phase boundary seems to be the most important study object of colloid chemistry (FREUNDLICH's "Kapillar-Chemie" may be considered as the most impressive milestone in this conception of colloid chemistry).

Eventually it was tried to regard the sols of hydrophilic colloids as two-phase systems. This resulted in the well known theory of KRUYT and BUNGENBERG DE JONG, illustrated by Fig. 1. But many facts became known which were not compatible with this hypothesis and today we may say that it has only a historical value.

A glance at the properties of three different types of sols will enable us to illustrate the fundamental criteria leading to the modern classification of colloids.

A. The gold sol consists essentially of very small gold particles. These particles are prevented from precipitation by their capillary electric charge. The addition of an indifferent salt diminishes the charge and the particles will precipitate *irreversibly.* The stability of the sol is in reality only an apparent one. Spontaneously every sol of this type will flocculate, though this may take several years. This spontaneous flocculation can be explained by the fact that a small (sometimes negligibly small) fraction of the particles will have sufficient kinetic energy to overcome the mutual electric repulsion. The most important characteristic of these types of sols (the *hydrophobic* sols) is: *the sol state is principially a non-equilibrium state.*

B. A protein sol may come into being spontaneously, when water is added to some protein crystals. In this solution the colloid particle is the

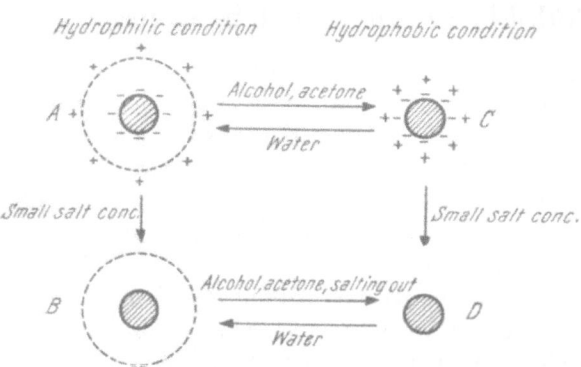

Fig. 1. Change of the kinetic units of hydrophilic colloids according to earlier ideas. *A* In pure water the kinetic unit is characterised by the presence of a double layer and a diffuse hydration. *B* After addition of a small amount of salt the double layer (not shown) has receded inside the hydration shell. *C* and *D* are comparable to *A* and *B* as regards charge, but the particles do not possess the diffuse hydration shell. *A* possesses two, *B* and *C* one, and *D* no stability factors.

protein molecule. The protein particles should be described as macromolecules in true solution in the water. Here *the sol state is the true equilibrium state.* The question whether a certain protein (or other macromolecule) is soluble in water depends—as in the case of small molecules—on its molecular structure. The number of charged and other hydrophilic groups is of the utmost importance as regards the solubility in water, vide the influence of pH on the solubility of proteins. From this point of view solutions of proteins and other biocolloids of macromolecular nature in water may be classified together with the solutions of macromolecules without electrolytic nature in e. g. benzene; the only important difference being the preponderating influence of hydrophilic groups in the former.

C. A soap solution does not always contain particles of colloidal dimensions. At very low concentrations (depending on the chain length of the soap) the soap ions are the kinetic units. At a certain concentration (critical concentration of micelle formation) the properties of the soap solution change abruptly. The reason is that at this concentration the soap ions unite into micelles, the charged groups pointing to the exterior of the micelle. So these sols bear a superficial resemblance to the hydrophobic sols: the colloidal particles consist of many associated smaller units. Thermodynamically speaking, however, the likeness to the macromolecular sols is much more important: the colloidal soap solution is formed spontaneously. Here too: *the sol state is the true equilibrium state.*

Thus we arrive at the following classification of sols:

A. The sol state represents principally a non-equilibrium state.

Hydrophobic (lyphobic) *sols*

B. The sol state represents principally an equilibrium state.

    a) The dissolved substance is present in true solution as single molecules.

    Macromolecular sols

    1. Macromolecules are non-electrolytes [1].

    2. Macromolecules are electrolytes [1].

    b) Only the reversible aggregates of the smallest possible kinetic units have colloidal dimensions.

    Sols of Association Colloids

The biologist will be interested practically exclusively in group B, so we will not deal with group A. In protoplasm we find macromolecules with electrolyte nature and association colloids.

The living cell is the seat of anabolic and katabolic processes, and moreover of processes which derive their energy from metabolic processes. If these energy producing reactions did proceed in a homogeneous medium, then the energy would be liberated in the form of heat. The living cell, however, could not function as a heat-engine, as the efficiency of a heat-engine depends on the difference of two temperatures in the engine $\left(\dfrac{T_1-T_2}{T_1}\right)$, whereas the temperature within the cell will be approximately equal. This emphasizes the thesis that the transformation of energy (e. g. in muscle the transformation of chemical into mechanical energy) must follow a more direct way. The living cell might be compared to a chemodynamic engine.

In its most simple form such an apparatus (galvanic cell) is characterized by a splitting up of the chemical reaction into reversible elementary processes and the separation of these partial processes in space. Thus we arrive at the conclusion that protoplasm—in order to be the seat of a chemodynamic device—should have a certain structure. The subdivision of protoplasm in several compartments (nucleus, plastids, mitochondria, cytoplasm, etc.) meets this demand more or less. One would presume, however, that especially submicroscopic elements of the protoplasmic structure will be of essential value to the chemodynamical processes. We will not yet ask how this protoplasmic structure should be represented, but only stress the idea that biocolloids will play a rôle of paramount importance in this structure.

We will now discuss the relation between the shape of the biocolloids and their function in protoplasm (and its derivatives). From this point of view the macromolecules might be classified into two categories:

    1. linear macromolecules.

    2. corpuscular macromolecules.

---

[1] The boundary between these two groups is far from sharp. An amylum molecule e. g. contains only a few charged groups. In some respects these groups are of great importance, so that one might classify this macromolecule as an electrolyte. In other respects the non-electrolyte character comes to the front.

It seems worth while to consider the ability of the macromolecules to fulfil certain functions in cell life. For comparison's sake we will also look at the possibilities of association colloids.

The *linear macromolecules* arise from a polymerisation of equal (e. g. glucose in cellulose) or unequal (e. g. amino acids in proteins) elements.

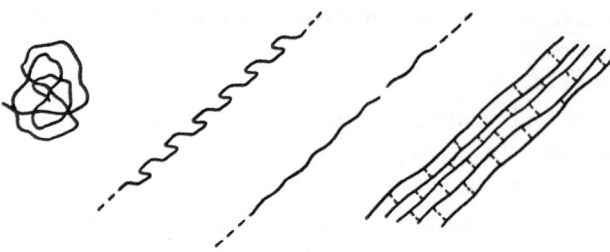

This long chain may have all kinds of shapes. In solution, without disturbing influences, it will practically always have the form of a more or less dense coil (or clew) kinked at random. The density of these coils fluctuates about a statistically most probable value.

Fig. 2. Structural possibilities of linear macromolecules.

In linear macromolecules with electrolytic nature the density of the coil is much influenced by the pH of the solvent. The diameter of the coil will increase with increasing charge and eventually the macromolecule will be completely stretched (Fig. 2). Here possibilities for contraction come to the fore. A bundle of closely packed linear macromolecules, united by strong (covalent) or many weak bonds (hydrogen bridges) may form a

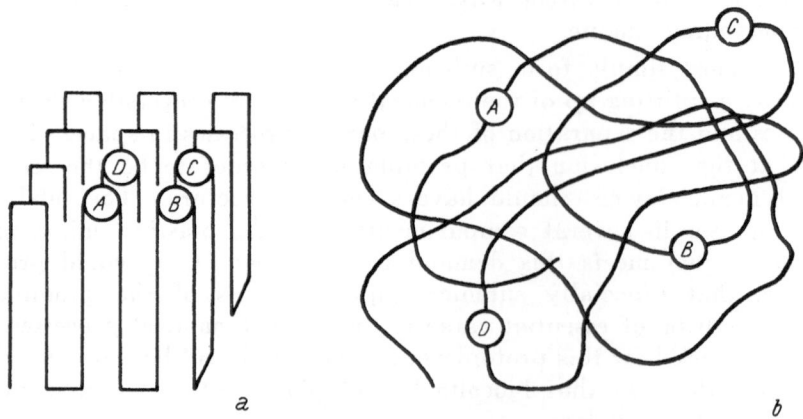

Fig. 3. Three-dimensional specificity of globular proteins. *a* native protein, *b* denatured protein.

structure of great strength. Thus we may derive from the shape and properties of linear macromolecules that we will find these biocolloids in:

a) supporting structures (e. g. cellulose fibrils in the cell wall),
b) contractile structures (e. g. myosin in muscle).

The *corpuscular (globular) macromolecules* (especially proteins) consist of linear chains folded in a specific way (Fig. 3). Of course, a linear protein has already a certain linear specifity as the amino acids are situated in the chain in a characteristic sequence. The specificity is increased in the globular proteins; the surface of such a protein corpuscle has a two-dimensional specifity. This surface provides a much greater possibility for specific binding and activating of other molecules than a linear macromolecule would have. The catalysts of the living cell will be globular proteins.

It is quite generally agreed that denaturation of globular proteins consists in an unfolding of the peptide chains (perhaps followed by an aspecific refolding). From Fig. 3 it may be easily deduced that the denaturation will practically always mean a loss of specificity and catalytic power. After denaturation several groups which were situated in the interior (and consequently could not be reached by certain chemicals) will become easily accessible.

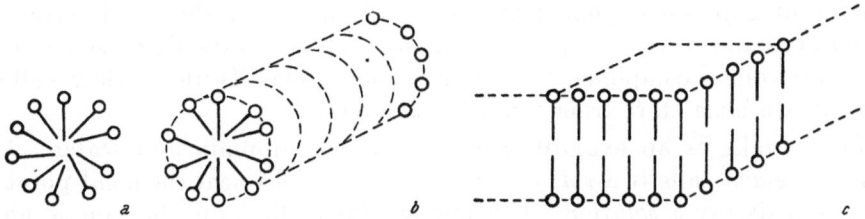

Fig. 4. Micelles of association colloids.

The *association colloids* show a variety of possibilities as regards the shape of the micelles (Fig. 4). Especially the large flat micelles (Fig. 4 c) merit our attention. Under certain circumstances they will grow rather large and will make contact here and there, thus giving the liquid a high viscosity and even an easily perceptible elasticity. This reminds one of the viscosity and elasticity of protoplasm.

The large micelles are in reality double films with a hydrophobic interior. Situated at interfaces, they may control the diffusion of small hydrophilic molecules. It has been surmised long ago that the outside layer of protoplasm, the myelin sheath of nerves, and many boundaries between cell compartments consist of orderly built layers of lipid material. In general we might say: if the living cell partitions off certain compartments, use is made of association colloids. On the other hand these colloids may (perhaps together with linear proteins) constitute the structural basis of protoplasm. The most remarkable properties of protoplasm (viscosity, elasticity and sol ⇄ gel transformation) might perhaps be traced back to micelles of association colloids.

References

Booij, H. L., 1949: Chapter XIV in Kruyt's Colloid Science II. Amsterdam.
Bungenberg de Jong, H. G., 1949: Chapter I in Kruyt's Colloid Science II. Amsterdam.
Overbeek, J. Th. G., and H. G. Bungenberg de Jong, 1949: Chapter VII in Kruyt's Colloid Science II. Amsterdam.

2. Colloid Systems

From the preceding section it has become clear that colloid chemistry is not a separate science, but that it should be seen as a part of physical chemistry. Perhaps it would be better to say two parts, as the hydrophobic sols and the colloid systems which represent equilibrium states (macromolecular and association colloids) are only very loosely bound together. The dimension of the kinetic units is the only important "theoretical" link. In practice, however, this weak link means some methodical agreement, which justifies the separation of colloid chemistry from physical chemistry to some extent.

The difference between a physical chemist and a colloid chemist would be—in theory at least—that a physical chemist is interested in all components of a poly-component system, while a colloid chemist is strongly biased in favour of the components having certain arbitrarily chosen dimensions (the colloid components). When trying to classify the various colloid systems, we must start from this predilection.

Let us take as an example a system of egg albumin and water. The colloid chemist calls it a *sol*, yet from a general physical chemical point of view it is simply a *solution*. We now mentally eliminate the kinetic units of all non-colloid components (thus all particles having a diameter below a certain value: "micro-units"). There remains a system of moving and colliding molecules, that on many points resembles a *gas*. Long ago Van 't Hoff discovered the "gas laws" in solutions of small molecules and colloid science recognised the validity of this principle in determining the molecular weight of colloids by osmotic methods, using these laws of dilute solutions.

Subsequently we take a solution of isoelectric gelatin (at 50⁰ C.) and add alcohol (or Na_2SO_4). The mixture remains clear up to a certain concentration. On further addition turbidity is produced. This turbidity is caused by a large number of minute drops, which coalesce to a viscous liquid layer (coacervate). This layer contains relatively much gelatin; in the other layer—equilibrium liquid—the gelatin concentration is very low. At a sufficiently high concentration cohering masses of floccules are obtained. After a long time these masses of floccules are also transformed into a coherent coacervate layer. Applying the same elimination procedure which we used in the case of the egg albumin solution, we here have two phases (a) solution of gelatin and (b) coacervate of gelatin, which show an analogy to a *gas* and a *liquid* respectively. In phase *a* we have few free moving kinetic units, in phase *b* the kinetic units are closely packed, but they are still in movement and they show *no three dimensional regularity*.

In the third place we point to the crystals which may be obtained from solutions of globular proteins by addition of $(NH_4)_2SO_4$ or other means. These crystals are rather variable (1) as they may contain micro-units (water, ions, etc.) in varying amounts and (2) as these micro-units may be replaced by others without a radical change of the crystal structure. From the colloid-chemical point of view they are homogenous; only the distance between the particles within the crystal may vary between certain limits. It is clear that these systems (which we will call *colloid crystals*) show an analogy to the normal crystalline solid. The corpuscular macromolecules are arranged in a three dimensional lattice.

Thus colloid science recognises three basic systems:

      I Sol (solution),   analogy gas
     II Coacervate,   analogy liquid
    III Colloid crystal,   analogy solid (cristalline).

Just as in the cases of the normal gases, liquids and solids, a coacervate will be in equilibrium with a sol (equilibrium liquid), in which the concentration of the colloid particle may be high or low (compare the equilibrium between a liquid and its vapour). The same applies to the equilibria between the other colloid phases.

Already in the usual classification of gases, liquids and solids many instances are known of systems which are not easily classifiable. It is not surprising that in colloid science many systems are found which a first sight do not fall into one of the three classes given above, or appear to be intermediates.

In the first place we may mention the liquid crystals. Theoretically speaking two intermediate possibilities exist between the isotropic coacervate and the colloid crystal with three dimensional regularity.

These possibilities have indeed been found: "paracrystals" in the nematic state (an array of anisodiametric molecules with a common axial direction and a random distribution, mobile in two spatial directions of centres

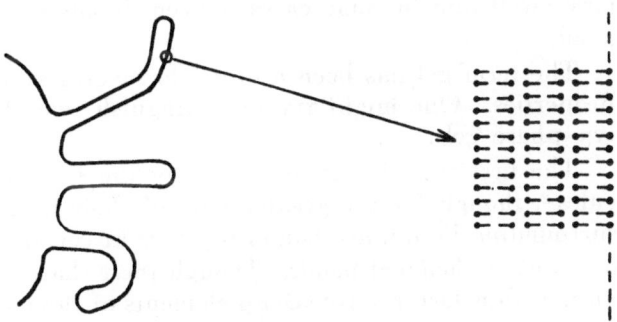

Fig. 5. Molecular structure of myelin tubes.

of gravity) and in the smectic state (the centres of gravity are mobile in one direction). There are instances known of a nematic colloidal phase (with relatively large amounts of water) in equilibrium with a—more concentrated—smectic phase. The well known myelin forms (originating from lecithin after the addition of water) fall into the category of paracrystals (smectic state). The X-ray diagrams show that these structures are built up from bimolecular layers (Fig. 5). It is interesting to note that the spacings characteristic for the bimolecular films increase continuously with increasing water content (from about $60\,\text{m}\mu$ to as much as $150\,\text{m}\mu$).

In many cases it is far from easy to distinguish to which class a colloid-rich phase belongs. Everyone knows that in trying to separate a colloid from a solution one often gets a mass of floccules. Only exceptionally the colloid will separate out in the form of a clear coacervate layer or a colloid crystal. It frequently costs a great deal of trouble to determine the nature of the "floccules" microscopically. They may be formed of minute highly viscous coacervate drops. Then gentle heating will sometimes produce larger drops which may be easily recognized. On the other hand they may be masses of cohering minute colloid crystals. Then identification with simple means is practically impossible.

The phases discussed up till now (sol, coacervate, colloid crystal and paracrystals) might be characterised as systems being macroscopically, microscopically as well as submicroscopically homogeneous. On the other hand many instances are known of colloid systems which are micro-scopically and macroscopically homogeneous, but submicroscopically in-homogeneous. We will give some examples:

a) *Inhomogeneous sols.* The whole system is liquid and has the character of a sol. It contains, however, submicroscopical aggregates which causes deviations of the normal properties of liquids (structural viscosity, sometimes even distinct elasticity).

b) *Inhomogeneous coacervates.* The coacervate shows the same devia-tions from the normal properties as were seen in the first case. Here too submicroscopical aggregates are the cause of the deviations.

c) *Gels.* The whole system is solid. When the solvent is added the gel may swell and in some cases it even dissolves. Then the gel has become a sol.

The term gel has been applied to several systems with rather different properties. One might try to distinguish two classes: one-phase gels and two-phase gels.

The first type is characterized by the fact that the single macromole-cules[2], though for the greater part of their length freely dispersed in the surrounding liquid, are bound together at certain points by cohesion forces or stronger chemical bonds. Though their character of independent kinetic units is thus lost, the free chain elements of the macromolecules still execute kinetic movements.

In its extreme form, this type of gel consists of macromolecules forming a coherent network throughout the whole system. One and the same macromolecule takes part in regions of the gel which might be called crystalline and in other regions, that resemble macromolecular solutions (Fig. 6). Thus there is formed a continuous lacunary system throughout the whole gel. It seems presumable that the one-phase concept is here more preferable than the two-phase concept. Still it is difficult to charac-terise the nature of this phase. The gel might for instance be considered as a crystalline phase with very extensive lattice disturbances. In a 2%

---

[2] or micelles of association colloids.

gelatin gel, the lattice disturbances would even amount to more than 98% of the total volume.

When we now suppose that the cohesion forces between the macromolecules are weak (and the time of contact short) we will get systems which show properties of solids as well as of liquids. We might then speak of easily deformable "solids" or of "liquids with structural viscosity and clearly visible elasticity". There is no sharp boundary between this type of gel and the inhomogeneous sols and coacervates.

Other gels, however, consist of a cohering mass of highly dispersed flocculation aggregates (the agar gel presumably belongs to this group). Here one is inclined to consider these gels as two-phase systems. For these two-phase gels special phase boundary considerations may be of use in explaining part of their properties. We approach the hydrophobic colloids. Of course, this type of gel is not as important for biology as the one-phase gel.

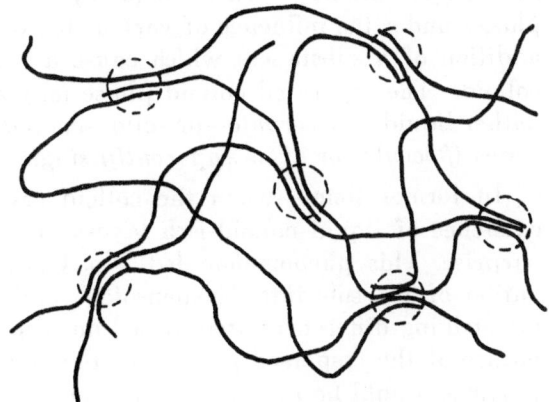

Fig. 6. Structure of a gel.

d) *Apparently single colloid systems.* We have seen that under certain circumstances insoluble inorganic substances like gold, sulfur, etc. may be subdivided to submicroscopical dimensions. The particles may form a more or less stable sol by virtue of the charged surfaces. It should be possible that in a macromolecular solution separation of a coacervate or a crystalline phase sets in, while this new colloid-rich phase remains subdivided in the range of colloidal dimensions. This system does not represent an equilibrium state, though it might at first sight give that impression, just like the "real" hydrophobic sols.

These systems do indeed exist in the field of macromolecular and association colloids. They behave like hydrophobic sols, they are flocculated [3] by neutral salts in small concentrations.

It will be clear that the "apparently single colloid systems" are closely related to the two-phase gels.

In the following chapters a survey of experiments on coacervates will take the greater part of the space available. Two reasons might be given for this preference. The first reason is that we have a wealth of experimental methods for studying coacervates in contra-distinction to the other colloid systems. The simplest method consists of measuring the coacervate

---

[3] Finally a single colloid-rich layer may sometimes be formed.

volume in relation to the total volume. Moreover one may measure the density, viscosity, refraction and similar properties. With other systems, e. g. colloid crystals or floccules, the number of experimental methods is far more limited.

The second reason is that of all colloid systems the coacervates resemble protoplasm most. This—perhaps superficial, perhaps important—resemblance has led to the study of *colloid morphology*. The important question whether or not protoplasm should be called a coacervate will be treated in one of the following chapters.

We will close this chapter with some general remarks on coacervates. As we have already seen a sol (one-phase system) may separate into two phases under the influence of various factors (change in temperature or pH, addition of a substance) which cause a reduction of the solubility of the colloid. The separated colloid phase may appear in a low dispersed state (either liquid—*coacervate*—or solid—*colloid crystals)* or in higher dispersed states (*floccules* or even *apparently single colloid systems*).

In former times—when the colloid crystals were not yet known—the existence of liquid colloid-rich layers in a two-phase system caused some surprise. This phenomenon led Wo. Ostwald to his well known classification of the sols into "suspensoids" and "emulsoids." He believed that the striking differences were based on respectively the solid and the liquid nature of the dispersed phase. In the gelatin-sol for instance, the protein particles would be present as ultramicroscopic liquid drops, as they would be united with a relatively large quantum of water. In this line of thought a visible separation into two liquid layers is only a change in the degree of dispersion of the second liquid phase already present in the emulsoid sol. Bungenberg de Jong and Kruyt, starting from much the same point of view, introduced the term *coacervation* for the phenomenon described.

Subsequently it became clear that coacervation and flocculation are very closely related phenomena and thus it was thought that coacervation too, might be explained by their stability theory (compare Fig. 1). The original sol particle was considered to be surrounded by a hydration coating of considerable size, in which the water was bound less and less tightly towards the periphery. Such a sol particle would owe its stability just to this diffuse character of the solvate coating, since the latter is not sharply defined at its periphery. Consequently it possesses no free surface energy. Transformation of the diffuse solvate coating into a sufficiently concrete outer boundary (consequently with free surface energy) will result in a union of the sol particles through their solvate coatings. As the actual particle nuclei are still displaceable with respect to each other the coacervate has the nature of a liquid (Fig. 7). This theory has served as a useful guide in the further experimental work, but its fundamental assumptions became more and more doubtful. One of the reasons to forsake this theory was the fact that the measurements of the amount of hydration of macromolecular biocolloids—though giving widely diverging results—never gave the enormous quantity of "bound water" required by this theory.

With the development of the modern theory of macromolecules came the conclusion that coacervation is a true partial miscibility in the sense of the phase theory. In this respect the phenomenon resembles the partial miscibility which occurs in systems consisting exclusively of micro-units (e. g. phenol/water, alcohol/water/$(NH_4)_2SO_4$, etc.). Nevertheless it is an extreme case of partial miscibility in so far as the two phases (coacervate and equilibrium liquid) differ practically only in colloid content (the con-

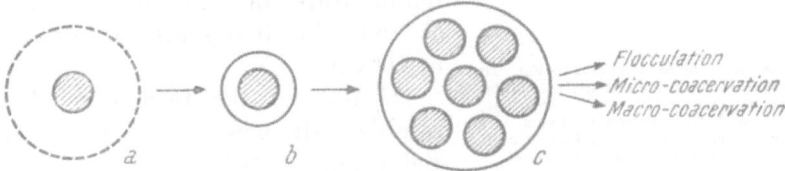

Fig. 7. Original scheme for the mechanism of coacervation.

centration of the colloid is very high in the coacervate and very low in the other phase). The concentration of the micro-units is more or less the same in both phases.

The concept of the statistically kinked macromolecule can be of service in explaining the facts on which the old coacervation theory was based. An example will make this clear.

1. When adding certain micromolecular substances (e. g. Na_2SO_4) to a dilute sol of a linear macromolecule (e. g. isoelectric gelatin) the viscosity $\left(\dfrac{\eta_{s} - \eta_{0}}{\eta_{0}}\right)$ decreases sharply in a certain concentration range of the added substance, previous to the coacervation.

2. On just exceeding the "coacervation limit" (the minimal concentration of Na_2SO_4 required to get coacervation), the coacervate still contains a relatively large amount of water and other micro-units. This amount decreases on further addition of the micromolecular substance.

The macromolecule is present in the form of a more or less dense coil. Addition of the micromolecular substance will decrease the solubility of the macromolecule, in other words, the affinity of various groups along the macromolecule for the solvent (water) decreases. Eventually it reaches the same value as the mutual affinity of the groups, and finally it decreases to lower values. Consequently the macromolecular coils grow much denser (lowering of viscosity) and some points of contact of more or less long duration are formed between the loops of different macromolecules. If this inter-molecular association is sufficiently great, coacervation takes place. Thus the fall of viscosity and coacervation are not the result of dehydration, but of the large reduction of the amount of occlusion liquid inside the macromolecule (see for a simple scheme Fig. 8).

Further the coacervate is to be regarded as an association of macromolecules (or micelles of association colloids) in which the points of contact are of a dynamic nature, since it is still a typical, though viscous, liquid.

Coacervation may be brought about in very different ways but we can give a classification into two large groups. These groups can be illustrated by two examples of micromolecular systems (it should always be remembered that coacervation results from a decrease in solubility).

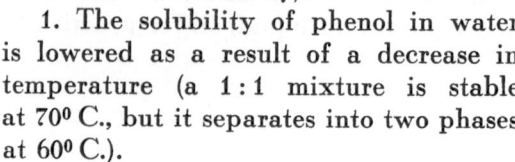

1. The solubility of phenol in water is lowered as a result of a decrease in temperature (a 1 : 1 mixture is stable at 70⁰ C., but it separates into two phases at 60⁰ C.).

Fig. 8. Modified scheme of coacervation. In the coacervate there are mutually associated macromolecules, which penetrate each other, at any rate with their peripheral loops. At the circumscribed volume of a loosely built macromolecular coil in the original sol.

2. We get the formation of an insoluble salt when mixing solutions of BaCl$_2$ and Na$_2$SO$_4$.

Thus we distinguish:

1. Simple coacervation; concerned with the non-ionised groups.

2. Complex coacervation; salt-bond formation, the charges on the macromolecules play the important part.

From a biological point of view the simple coacervation (e. g. gelatin with alcohol, resorcinol or Na$_2$SO$_4$) is not very interesting, so we will turn our attention to the second type of coacervation. Before doing so we must discuss the experiments on the charge of macromolecules.

References

Bungenberg de Jong, H. G., 1949: Chapter VII in Kruyt's Colloid Science II. Amsterdam.

3. Colloids with Electrolytic Nature

The difference between the electric charges of hydrophobic colloids and macromolecular colloids is that the former derive their charge from ions adsorbed to the particles, in the latter it is due to the dissociation of groups firmly attached to the macromolecule. Thus the macromolecular and association colloids with electrolyte character can be divided into:

a) Colloids with acid character, which carry only anionic groups such as —COO⁻, —OSO$_3$⁻ and —OPO$_3$H⁻.

b) Colloids with basic character, which carry only cationic groups, such as —NH$_3$⁺, —NHC(NH$_2$)$_2$⁺ and —N⁺(CH$_3$)$_3$.

c) Colloids with amphoteric character, which carry both types of groups.

When looking for a quantitative explanation of the way in which the composition of the medium determines the charge of the macromolecular electrolytes, one should use the theories of electrolytic dissociation rather than the theory of the double layer. Thus gum arabic will loose its negative charge at pH = 2 as the —COOH group acts like a rather weak acid

(Fig. 9). In proteins it is sometimes possible to analyse the titration curves and to attribute the various steps in these curves to the different types of charge-carrying groups. We will not enter into the difficulties encountered in this method, but restrict ourselves to the examples of macromolecules carrying only one type of group.

The most useful tool for the investigation of the charge of colloids is the electrophoresis. It has an advantage over the titration method as it enables one to measure not only the influence of H+ and OH⁻ ions but the influence of other ions as well. Furthermore it is frequently difficult to determine the charge-zero with the titration method, while this can be easily done with electrophoresis.

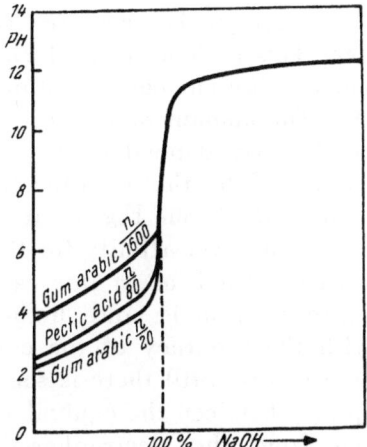

Fig. 9. Titration curves of gum arabic and pectic acid.

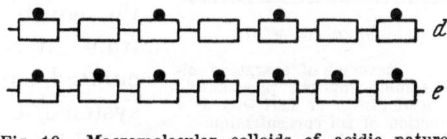

Fig. 10. Macromolecular colloids of acidic nature arranged in a series of increasing charge density (decreasing equivalent weight). Rectangles: ☐ monomeric residues; black dots: ● negative groups.

It will be clear that the number of charges per unit weight of the macromolecule is a very important characteristic (density of charge). In Fig. 10 we pictured a number of macromolecules of the same type, but with an increasing number of ionising groups. In accordance with our efforts to use—if possible—normal physical chemical terms, we will describe the charge of macromolecules in terms of equivalent weight. Of course, the lower the equivalent weight, the higher the charge density.

To determine the equivalent weights of a number of biocolloids TEUNISSEN and BUNGENBERG DE JONG studied the influence of hexolnitrate [4] on the electrophoretic velocity of these colloids. The hexavalent cation has a strong affinity for negative groups situated on colloids. Thus in increasing the concentration the negative charge of the colloid grows less, reaches the value zero, while at still higher concentrations the charge becomes positive. The modified apparatus of NORTHROP is very suitable for measuring the electrophoretic velocity (which depends on the charge). As most colloids are soluble (and thus invisible in the microscope) a suspension of very small quartz particles is added to the mixture to be measured. Each particle is—at suitable colloid concentrations—covered

---

[4] = [Co {(OH)$_2$ Co (H$_2$N—CH$_2$—CH$_2$—NH$_2$)$_2$}$_3$] (NO$_3$)$_6$

with a colloid film and the whole now behaves as a colloid particle. The electrophoretic velocity at several concentrations of hexolnitrate is plotted against the concentration of phosphatide. From the graph obtained we read by interpolation the concentration of hexolnitrate needed to reach zero charge. This concentration at zero charge consists of two elements: the amount of hexol-ions required to neutralise the negative charges of the colloid plus a certain equilibrium concentration. As we are only interested in the former, it is necessary to eliminate the latter. This is done by performing the experiment at different concentrations of colloid (Fig. 11). The amount of hexol-ions bound is of course linearly dependent on the colloid concentration, while the equilibrium concentration is constant. From Fig. 11 it is seen that an ion having a lower affinity for the negative groups is less suited, as in this case the equilibrium concentration is much higher (this interferes with the accuracy of the calculated equivalent weight). Still there is some systematic discrepancy between the equivalent weights calculated from the electrophoretic measurements with hexolnitrate (which are called *reciprocal hexol numbers*) and the analytically determined equivalent weights, the former being about 16% lower than the latter[5]. Table 1 gives an idea of the values

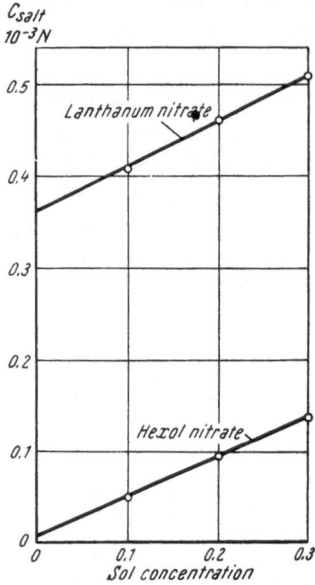

Fig. 11. Reversal of charge of alcohol soluble soybean phosphatide with hexolnitrate or La(NO₃)₃ as a function of sol concentration.

Table 1.

Substance	Reciprocal hexol number	Equivalent weight
Na-pectate	203	233
Na-carrageen	233	294
K-chondroitin sulfate . .	290	319
Na-yeast nucleinate . .	294	373
Soybean phosphatide . .	782	988
Na-pectinate	1,040	1157
Na-arabinate	1,068	1202
Na-agar	2,264	2909
Soluble starch	26,000	—
Glycogen	78,000	—

determined in some biocolloids. It is clear that in most cases the reciprocal hexol numbers give a sufficient approximation of the "charge density."

---

[5] For an explanation of these differences see Bungenberg de Jong in Kruyt's Colloid Science II. Amsterdam 1949, p. 266.

We included in Table 1 an association colloid (soybean phosphatide). At first sight one would expect that such lecithin-like molecules would have a charge density zero (or equivalent weight ∞) as the number of positive groups equals that of the negative ones. Crude phosphatide preparations, however, always show a negative charge. This charge may presumably be ascribed to a small admixture of phosphatidic acid. Another result of the presence of phosphatidic acid is the fact that the iso-electric point of most lecithin preparations lies very low (for pure lecithin the isoelectric point should be at about pH = 7, or, more precisely, there should be an isoelectric zone from about pH = 4.5–9.5).

In a semi-quantitative way it is possible to calculate the proportion of phosphatidic acid to neutral phosphatides from the "equivalent weight." When a phosphatide is replaced by a phosphatidic acid we get a divalent ion instead of an ampholyt. Suppose we find in a mixture of n molecules lecithin one molecule phosphatidic acid (Fig. 12). The molecular weight of lecithin will be in the neighbourhood of 780, that of phosphatidic acid will be approximately 680. Then the "equivalent weight" of this mixture will be:

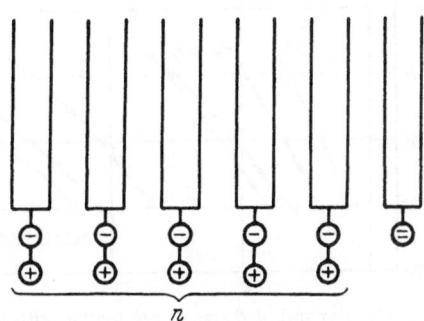

Fig. 12. Lecithin as it occurs in nature is negatively charged as it contains some phosphatidic acid.

$$\frac{n \times 780 + 680}{2} = 390\,n + 340.$$

In the case of the soybean phosphatide cited in Table 1 we find $n = 1.7$, which means that the ratio between phosphatides and phosphatidic acids will be about 17 : 10. It is clear that in these cases the word equivalent weight has not much sense. Still we have to remember that the association colloids practically always form micelles. As we are dealing with mixed micelles we might say that the "individuality" of the components disappears and that we may look upon these micelles as units with a certain "equivalent weight." This will be permitted only if the micelle is composed of a large number of molecules.

Beside the equivalent weights a second characteristic determines the behaviour of colloids. The negatively charged biocolloids may be divided into three classes, according to the composition of the ionised groups. This comes to the fore already in the different flocculability under the influence of added cations. Thus, at the same "charge density," the flocculability increases in the order:

sulfate colloid < carboxyl colloid < phosphate colloid.

These differences are still more conspicuous when one measures the influence of cations on the charge of the biocolloids. In the first place we may point to the different affinities of H-ions to the ionised groups. Thus "carboxyl colloids" will lose their negative charges relatively easily by lowering the pH. "Phosphate colloids" will lose their charge at a much lower pH. The "sulfate colloids" have ionised groups with a still stronger acid nature. Secondly, the ionised groups have different polarisabilities, which fact will play an important rôle in the binding of inorganic ions on these groups.

When studying the reversal of charge of colloids specific cation sequences come to the fore. The method used has been described already (see page 5).

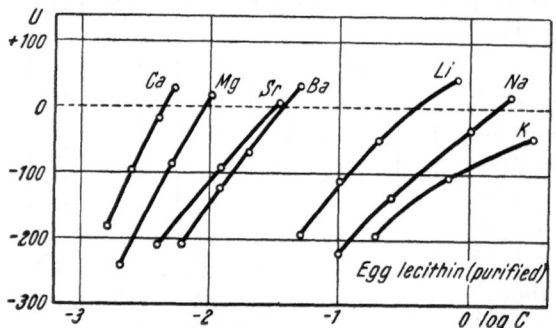

Fig. 13. Reversal of charge of egg lecithin with alkaline and alkaline earth chlorides.

We must remember that the experimentally determined reversal of charge concentration of a certain cation is composed of two parts:

a) a true reversal of charge concentration (C_t = the concentration of the cation in the medium at the reversal of charge point),

b) a fictitious concentration (C_f = representing the amount of cations bound to the colloid; and thus depending a. o. on the concentration of the colloid).

The concentration found experimentally will be the sum of the concentrations mentioned:

$$C = C_t + C_f.$$

In determining the equivalent weight we were interested in C_f, now will be interested in C_t (the equilibrium concentration in Fig. 11). In the case of mono- and divalent ions (and at small concentrations of colloid) our formula approximates to $C = C_t$, as C_f is very small compared to C_t. Only in the case of ions with a strong affinity for the charged groups of the colloid we will have to correct the concentrations at zero charge experimentally found.

If we are not interested in the absolute values of C_t—in other words if we are only interested in the sequence of the cations as regards their affinities to the colloid—we may use the measured concentrations directly. This is permitted as at a certain sol concentration C_f will be equal for all ions.

In Fig. 13 we give the results of a series of experiments on the influence of cations on the electrophoretic mobility of egg lecithin (purified via the $CdCl_2$ compound; measured on suspended quartz particles).

From this graph we read by interpolation the concentration at zero charge. In the following diagrams (Fig. 14) we have collected the so-called "reversal of charge spectra" of some phosphate colloids.

We draw the attention to the following points:

1. The pronounced influence of the valency of the cation. Compare the groups Li-Na-K, Mg-Ca-Sr-Ba, and Ce-La.

2. Increasing the ion radius in the group Li-Na-K and Ce-La leads to increase of the reversal of charge concentration. In the case of Mg-Ca-Sr-Ba irregular series occur.

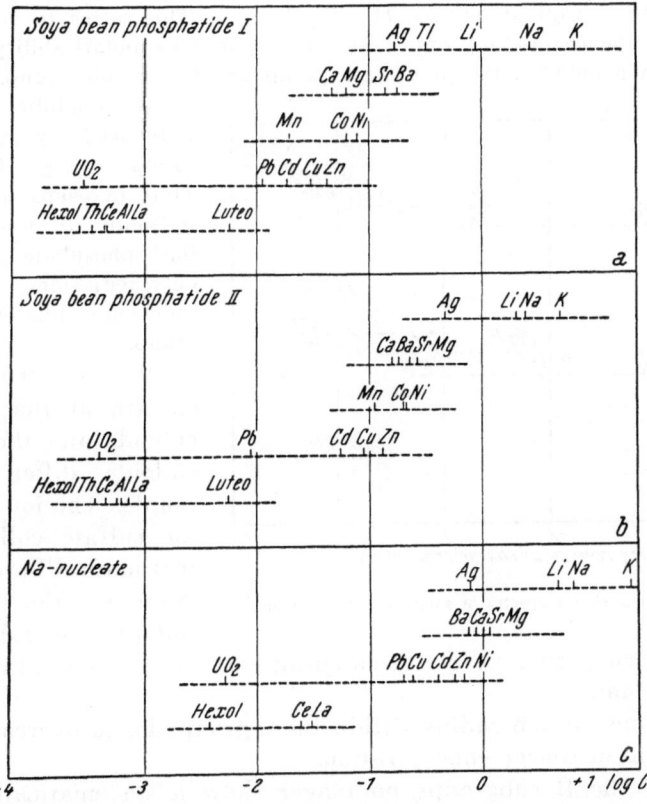

Fig. 14. Reversal of charge spectra of some phosphate colloids. *a* Alcohol soluble soybean phosphatide, "sensitised" with triolein. *b* Alcohol insoluble soybean phosphatide. *c* Na-nucleate.

3. Ions of the B subgroups of the periodic system show generally smaller reversal of charge concentrations than those of the A subgroups with equal valency. Compare Ag and Tl with Li, Na, and K; or Pb, Cd, Cu and Zn with Mg, Ca, Sr, and Ba.

4. UO_2, though only divalent, shows an exceptionally low reversal of charge concentration.

We will compare these data with the ion spectra of some carboxyl colloids (Fig. 15).

Here the following points are of interest.

1. The influence of the valency of the cations in the A subgroups of the periodic system is still present, but it is less pronounced than in the case of the phosphate colloids.

2*

2. Increasing the ion radius in these subgroups has the reverse effect (the trivalent ions Ce and La excepted).

3. Ions of the B subgroups show in general smaller reversal of charge concentrations than those of the A subgroups (with the same valency).

4. UO_2 occupies no exceptional place.

It must be remarked that all carboxyl colloids quoted are derived from carbohydrates. The cation sequences for other carboxyl groups (soaps, proteins) may be different (Colloid Science II, chapter IX). This seems to be caused by the fact that the carboxyl group takes—as regards its polarisability—an intermediate position between the phosphate group and the sulfate group. Moreover, its polarisability is strongly influenced by neighbouring groups (e. g. OH-groups). Thus in soaps and proteins it more or less resembles the phosphate group; in polysaccharides it behaves somewhat like the sulfate group.

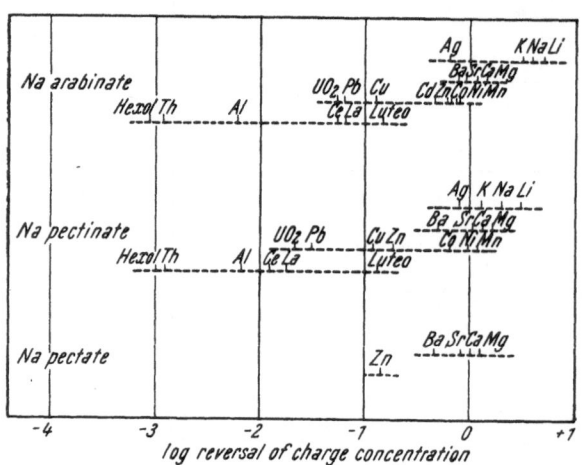

Fig. 15.   Reversal of charge spectra of some carboxyl colloids.

We see that the ion spectra of the phosphate colloids and the carboxyl colloids differ in some points. The ion spectra of the sulfate colloids show still more differences with those of the phosphate colloids (Fig. 16).

1. The valency (within the A subgroups of the periodic system) is no longer important.

2. Increasing the ion radius within these groups has a decreasing effect on the reversal of charge concentration.

3. Ions of the B subgroups no longer show a systematically smaller reversal of charge concentration than those of the A subgroups.

4. The UO_2-ion does not show an exceptionally low reversal of charge concentration.

The difference between the three types of colloids have been attributed to differences in polarisability of the ionised group. Bungenberg de Jong assumed the following order of polarisability:

$$\text{phosphate group} > \text{carboxyl group} > H_2O > \text{sulfate group}$$

We will not enter into a discussion of this hypothesis, but only state that the affinity between inorganic ions and ionised groups of colloids will depend on *valency, radius* and *polarising power* of the inorganic ion, on the *polarisability of the ionised group* of the colloid and—as the process takes place in aqueous medium—on the *polarisability of water*.

For a biologist the most important fact is the difference between the ion spectra of the three types of negative biocolloids. This fact leads to a new type of biological investigation. One might compare the "ion spectrum" of a certain biological process to the known ion spectra of colloids. This permits us to indicate which type of colloid is involved in the biological process.

In literature (see e. g. Höber 1945) one encounters many examples of investigations on the influence of cations on biological processes. Very often irregular series come to the fore, e. g.

$$Li < Cs < Rb < Na < K \text{ and}$$
$$Ca < Mg < Sr < Ba.$$

In colloid chemistry (reversal of charge of biocolloids) we meet these irregular series especially when working with phosphate colloids. It must be concluded that in many biological processes phosphate colloids play an important rôle.

The background of these irregular series has been elucidated by Bungenberg de Jong (1949). These series must be regarded as transition sequences between two extremes, e. g.:

$$Li \rightarrow Na \rightarrow K \rightarrow Rb \rightarrow Cs$$
$$Li \rightarrow Na \rightarrow K$$
$$Cs \leftarrow Rb \leftarrow \rfloor$$

$$Cs \rightarrow Rb \rightarrow K \rightarrow Na \rightarrow Li.$$

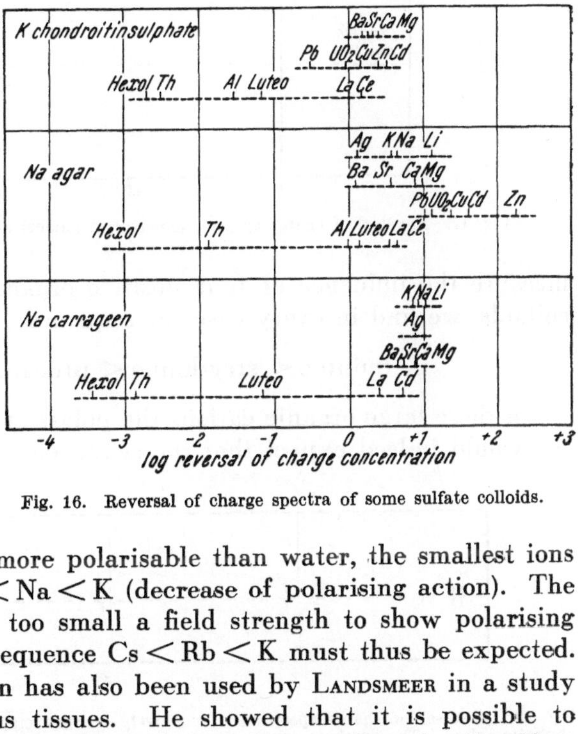

Fig. 16. Reversal of charge spectra of some sulfate colloids.

As the phosphate group is more polarisable than water, the smallest ions will show the sequence $Li < Na < K$ (decrease of polarising action). The large ions Cs and Rb have too small a field strength to show polarising action. For these ions the sequence $Cs < Rb < K$ must thus be expected.

This type of investigation has also been used by Landsmeer in a study of metachromasia in various tissues. He showed that it is possible to distinguish between the metachromasia of phosphate, carboxyl and sulfate colloids by the addition of some selected salt at various concentrations. Then the ion spectra of the depressing activity on metachromasia come to the fore.

The reversal of charge of positive groups by anions is given in Fig. 17. Here too we see that within a group of monovalent ions a certain sequence comes to the fore. Moreover it has been demonstrated (e. g. in the case of the reversal of charge of casein at pH = 3.4) that the valency of the ions is a factor of primary importance. Teunissen-van Zijp (1938) found the sequence (increasing reversal of charge concentrations):

$$K_4Fe(CN)_6 < K_3Fe(CN)_6 < K_2SO_4 < KCl.$$

As regards the reversal of charge of biocolloids (negative as well as positive) by organic ions, the first important fact is that the various types of colloids (phosphate, carboxyl and sulfate) can not be characterised by differences in sequence of a certain number of organic cations. When we

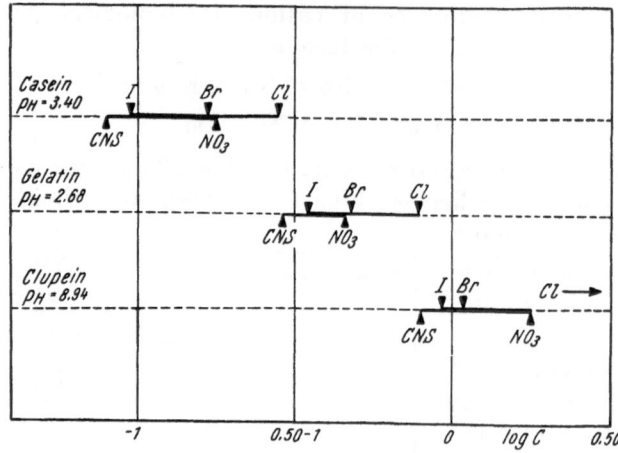

Fig. 17. Reversal of charge spectra (monovalent anions) of some positively charged proteins.

measure the influence of four positive organic ions on the three types of colloids, we find in every case the sequence:

quinine < strychnine < procaine < guanidine.

In these large organic cations the polarisation power is absent, and thus we would indeed expect that the sequence will always be the same. The

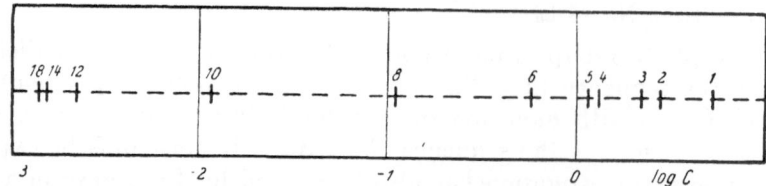

Fig. 18. Reversal of charge spectrum of positively charged clupein (pH = 8.94) with monovalent normal aliphatic anions. The numbers denote the number of carbon atoms in the chain (e. g. 1 = formate, 2 = acetate, etc.).

large differences in affinity must be ascribed to the differences in the hydrophobia of the organic cations. The larger the hydrophobic part of the ions, the lower the reversal of charge concentration. This is clearly demonstrated in Fig. 18, where the concentrations of a homologous series of monovalent ions, needed for the reversal of charge of the positive protein clupein, have been given. In the case of very long aliphatic ions (e. g. stearate) the equilibrium concentration (C_t) is very low, the "concentration" measured is in reality equivalent to the amount of ions bound to the protein (C_f).

It is interesting to note that acetylcholine and adrenaline show in certain physiological conditions K- and Ca-like actions. Working with the more stable ephedrine instead of adrenaline TEUNISSEN-VAN ZIJP (1938) showed that only as regards the phosphatides an influence comparable to the action of K and Ca is found for acetylcholine and ephedrine. Neither sulfate colloids nor carboxyl colloids (nor even nucleate) show this remarkable resemblance. The experiments give strong argument for the view that adrenaline and acetylcholine exert their Ca- and K-like action on phosphatides. Moreover, they point to the idea that the organism sometimes uses organic cations instead of inorganic ones as the former may be eliminated quickly by metabolic processes, which is of course impossible for the latter.

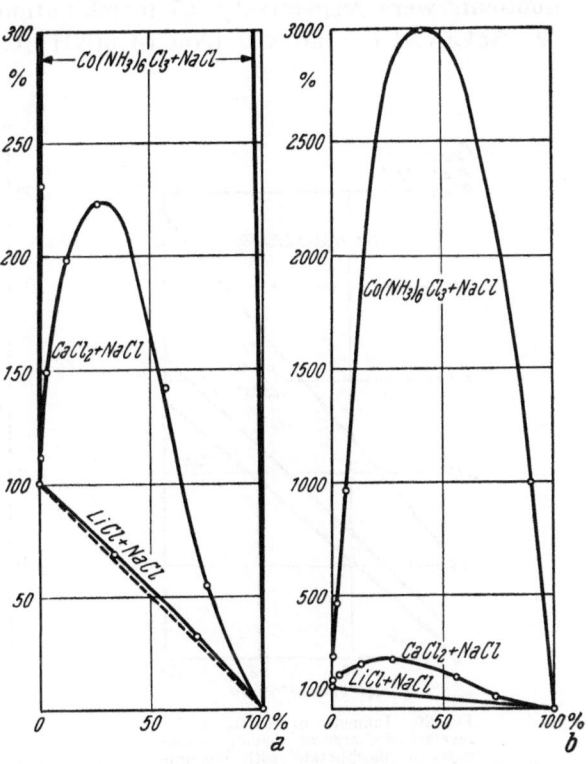

Antagonism of ions

An interesting phenomenon in physiology is the fact that some cations may antagonise the action of others. As the primary action of cations will be a decrease of the negative charge of biocolloids, it seemed worth while to look for antagonism of ions in vitro.

Fig. 19. Reversal of charge of alcohol soluble soybean phosphatide with mixtures of LiCl + NaCl, CaCl₂ + NaCl and Co(NH₃)₆Cl₃ + NaCl. Ordinates: concentrations of LiCl, CaCl₂ or Co(NH₃)₆Cl₃ expressed in per cent of the reversal of charge concentration of these salts in the absence of NaCl. Abscissae: concentration of NaCl in the salt mixtures expressed in per cent of the reversal of charge concentration of NaCl alone. The curves give the points where lecithin shows a zero charge. Above the curve the charge is positive, below it is negative.

BUNGENBERG DE JONG, BOOIJ, and WAKKIE (1936) studied the reversal of charge of biocolloids with salt mixtures. Fig. 19 gives some of the results obtained with a phosphatide. Here the reversal of charge concentration of each salt separately is given as 100% (NaCl is always plotted along the abscissa). In the case of the combination LiCl-NaCl we see that there is only a slight deviation from additivity. In the mixtures CaCl₂-NaCl a curious phenomenon comes to the fore. The reversal of charge concentration of CaCl₂ alone is 0.045 n. In the presence of about 1 n NaCl we need more than twice this amount (the reversal of charge concentration of NaCl alone is 2.8 n). Thus a large amount of NaCl antagonises the action of CaCl₂. This phenomenon is very pronounced in a mixture of a trivalent and a univalent ion $(Co(NH_3)_6Cl_3$-NaCl). Here we need in the presence of $1\,n$ NaCl a $Co(NH_3)_6Cl_3$ concentration which is about 30 times as high as the blank.

A further study—with other biocolloids—revealed that ion antagonism occurs when the quotient (Q) of the reversal of charge concentrations of the salts separately is larger than 10. In the cases depicted in Fig. 19 the quotients were respectively 2.5 (combination NaCl-LiCl; no antagonism), 62 (NaCl-CaCl$_2$) and 400 (NaCl-Co(NH$_3$)$_6$Cl$_3$). In Na-arabinate, on the

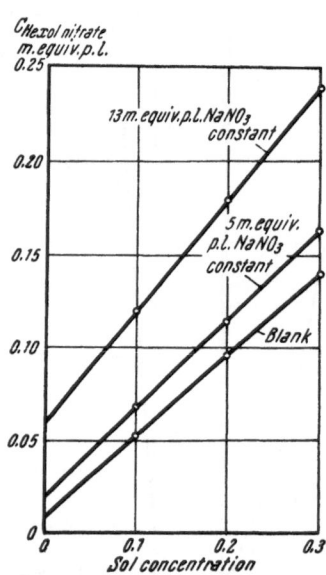

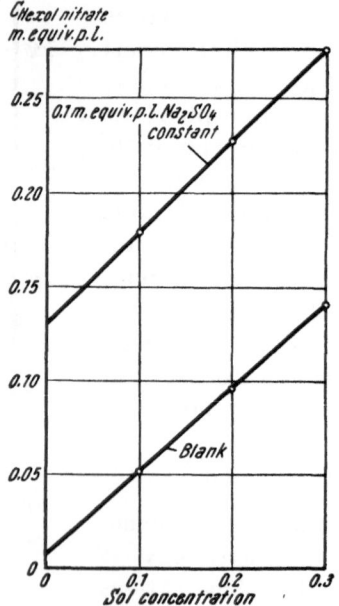

Fig. 20. Influence of NaNO$_3$ on the reversal of charge of alcohol soluble soybean phosphatide with hexolnitrate as a function of sol concentration.

Fig. 21. Influence of Na$_2$SO$_4$ on the reversal of charge of soybean phosphatide with hexolnitrate (compare Fig. 20).

other hand, the quotient (combination NaCl-CaCl$_2$) is 4.3. There no antagonism between NaCl and CaCl$_2$ has been found. From these considerations it is clear that the ion spectra (Fig. 14, 15 and 16) give the necessary information about the question whether antagonism will be found between two cations as regards the reversal of charge of a certain biocolloid. The ion spectra of phosphatides show the greatest spread of the cations, which means that with these biocolloids antagonism of the ions will often be found. The phosphatides are the only biocolloids in which antagonism can be readily observed in the combination NaCl and CaCl$_2$. Moreover, the purer the lecithin preparation, the larger the value of Q and consequently the more pronounced the antagonism will be. This seems to indicate that in those biological phenomena where antagonism between CaCl$_2$ and univalent cations has been found, phosphatides will be involved in the process.

It has been shown that the anions play an important rôle in this so-called cation antagonism. This view resulted from an investigation on the influence of $NaNO_3$ and Na_2SO_4 on the reversal of charge of phosphatides by hexolnitrate. Fig. 20 shows the concentration of hexolnitrate needed for the reversal of charge of alcohol soluble soybean phosphatide at three concentrations of the colloid (compare Fig. 11). The addition of $NaNO_3$ causes a shift of the curve. From the diagram it may be read that there is a strong increase of the true reversal of charge concentration (equilibrium concentration), while the amount of hexol-ions bound per gram colloid (indicated by the slope of the line) is not altered very much. For instance in the presence of 13 m. eq. p. l. $NaNO_3$ the true reversal of charge concentration is increased 7.9 fold, whereas the amount of bound hexol-ions is increased only 1.37 times. Two factors are involved: 1) the excess of negative ions lowers the "activity" of the hexol-ions and 2) every hexol-ion attached to the colloid is a point of attraction for anions.

In the case of Na_2SO_4 (Fig. 21) even a very small amount produces a very strong effect. Here the true reversal of charge concentration increases very considerably, while the amount of fixed hexol-ions is raised only slightly. Replacing the monovalent NO_3-ion by the divalent SO_4-ion causes a much stronger antagonistic effect. This comparison clearly shows that the denomination "cation-antagonism" cannot be an adequate one. The mechanism of this antagonism must be sought not in the fixation points (negative groups) of the biocolloid, but in the surrounding medium. The added anions will diminish the activity coefficient of the hexol-ions in the medium. Then divalent anions will have a much stronger influence than monovalent ones.

References

BUNGENBERG DE JONG, H. G., 1949: Chapter IX in KRUYT's Colloid Science II. Amsterdam.
— H. L. BOOIJ and J. G. WAKKIE, 1936: Zum Mechanismus des in Gemischen von Neutralsalzen auftretenden Antagonismus hinsichtlich der Umladung von Phosphatiden. Kolloid-Beih. 44, 254—284.
— und P. H. TEUNISSEN, 1938: Negative, nicht amphotere Biokolloide als hochmolekulare Elektrolyte I. Kolloid-Beih. 47, 254—320.
HÖBER, R., 1945: Physical Chemistry of Cells and Tissues. Philadelphia, Toronto.
LANDSMEER, J. M. F., 1951: Influence of pH and salts on metachromatic phenomena evoked by toluidine blue in animal tissue. Acta Physiol. Pharmacol. Neerl. 2, 112—128.
TEUNISSEN, P. H., 1936: Lyophiele niet amphotere bio-kolloiden beschouwd als electrolyten. Thesis, Leiden.
— und H. G. BUNGENBERG DE JONG, 1938: Negative, nicht amphotere Biokolloide als hochmolekulare Elektrolyte II. Kolloid-Beih. 48, 33—92.
TEUNISSEN-VAN ZIJP, 1938: Invloed van anorganische en organische ionen op de lading van biokolloiden, in het bijzonder van eiwitten. Thesis, Leiden.

4. Complex Systems

In chapter 2 we met two types of coacervation, the simple coacervation and the complex coacervation. The latter resembles the salt formation of inorganic chemistry. When we speak of *complex-relations*, we mean the electrical attraction between oppositely charged groups. Thus a complex-coacervate is a coacervate resulting from the interaction of two oppositely charged components (one at least should be a colloid). Of course, these complex relations may be present in all kinds of colloid systems (one might imagine complex sols, complex gels, complex coacervates, and so on).

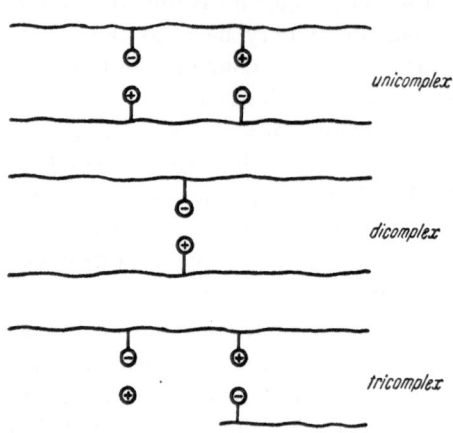

Fig. 22. Types of complex systems.

A rational classification of the complex systems ought to rest on the type of ions between which the complex-relations exist. As we know three types of ions (cations, anions and amphions) we classify the complex systems according to the number of types of ions (Fig. 22). If the ions are of only one type we speak of a *unicomplex system* (these ions must then be amphions). In the case of two ions we speak of *dicomplex systems*. Finally in some cases three ions are involved (amphion, cation and anion) and then we call them *tricomplex systems*.

As a further indication we may also state the nature of the system. Thus we get a binary nomenclature (examples: a unicomplex gel, a dicomplex coacervate, a tricomplex flocculation, etc.). Many variants of the three principal types are possible. To give one instance: a dicomplex system might consist of a colloid cation and a colloid anion or of a colloid cation and a micro anion or of a micro cation and a colloid anion [6].

a) Unicomplex systems

At its iso-electric point, a protein will not be really uncharged, but the number of positive and negative groups are equal. Thus it must be assumed that at or in the immediate neighbourhood of the isoelectric point complex relations will exist between the ionised groups. Sometimes (in the case of isolabile proteins) the proteins will separate out of the solution, in other cases (isostable proteins) the solution will remain clear at the isoelectric point.

Sols of the latter type (e. g. a gelatin sol) can be regarded as *unicomplex sols*. At a relatively high concentration of gelatin and at room temperature we get a *unicomplex gel*.

One of the most striking properties of complex systems in general is

---

[6] The limiting case—which is no longer a colloid system—consists of a micro cation and a micro anion. Several examples have been found of mixtures of solutions of micro cations and anions which give a separation into two liquid layers.

the fact that neutral salts weaken the complex-relations. Under the influence of a sufficient concentration of NaCl e. g., a complex coacervate or flocculation may disappear altogether. The valency of the ions plays an important rôle in this "solvent action," the higher the valency of the ion, the stronger the solvent action. This can be given in the following rules ("double valency rule"): the solvent action of neutral salts decreases in the series

$$3\text{--}1 > 2\text{--}1 > 1\text{--}1$$
$$1\text{--}3 > 1\text{--}2 > 1\text{--}1 \,[7].$$

Some arguments favour the view that e. g. a gelatin sol at its iso-electric point may be seen as a unicomplex sol. The relative viscosity

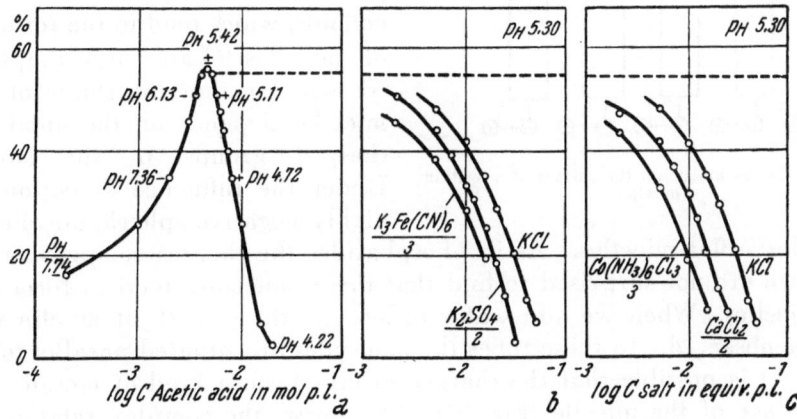

Fig. 23. Flocculation of serum globulin (horse) in the neighbourhood of the isoelectric point (a) and suppressive action of salts (b and c) in the vicinity of the maximum of curve a. Ordinates: turbidity.

increases after the addition of small amounts of neutral salts. The valency of the ions is important; the rules given above have been found for this phenomenon. The increase of viscosity can be conceived as a suppression of the complex relations (the density of the protein coils diminishes as the attraction of the charged groups is decreased).

Another experiment points in the same direction. An iso-electric gelatin sol is made slightly opalescent by the addition of a certain amount of alcohol. This opalescence disappears on addition of a small amount of a neutral salt.

Experiments on unicomplex flocculation are still more convincing. Globulins belong to the isolabile proteins; they flocculate at certain values of pH (Fig. 23 a). This flocculation disappears at relatively low concentrations of added salts (Fig. 23 b and c). Here we find the well-known properties of isolabile proteins: they dissolve in dilute salt concentrations. It is important to note that the double valency rule clearly comes to the

---

[7] Where 3—1 denotes a salt with a trivalent cation and a monovalent anion, 1—3 a salt with a monovalent cation and a trivalent anion, etc.

fore. This indicates that the isoelectric flocculation is a result of the complex relations existing between the charged groups of the protein. In some cases—the isolabile proteins—these forces are strong enough to cause a visible flocculation, in other cases—the isostable proteins—these forces are not large enough to surpass the solubility.

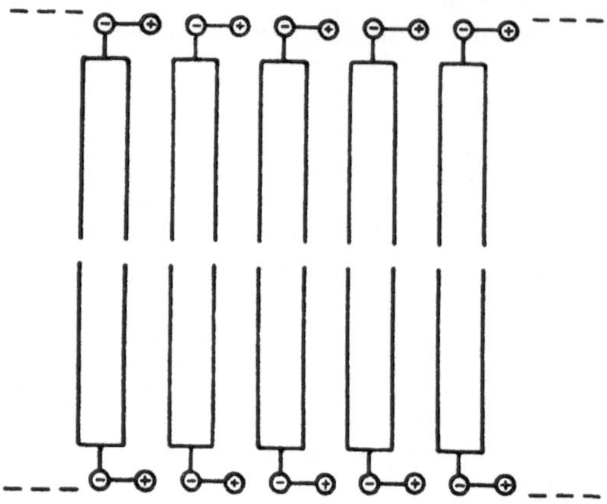

Fig. 24.   Charge mosaic on the surface of a lecithin micelle.

The complex relations must play an important part in pure lecithins and cephalins. In these molecules —at least at certain pH-zones—the number of positive and negative groups is exactly equal. The phosphatides belong to the association colloids, which tend to the formation of micelles. Later (see chapter 5) we will see that the shape of soap micelles depends on the number of charged groups in the surface. Under the influence of cations the highly negative spheric micelles become large flat micelles. As in phosphatides the charges compensate each other we are not surprised to find that these molecules tend to form large flat micelles. When we add water to lecithin the growth of myelin tubes may be observed. In these tubes the molecules are situated parallel to each other. It is possible that the charged groups form a kind of mosaic along the surface of the micelle (Fig. 24). Of course, the complex relations between these charged groups will be influenced by salt (see chapter 8).

b) Dicomplex systems

We may distinguish two types of dicomplex systems. Firstly there is the system resulting from the interaction of a colloid anion and a colloid cation. In the second place we may point to the systems consisting of a colloid ion and a micro ion (colloid cation + micro anion or colloid anion + micro cation). We will first pay attention to the interaction of two oppositely charged colloids.

The best known example in this realm is the system formed from gelatin and gum arabic (the latter substance, when purified and provided with Na-ions only, will be called Na-arabinate). We will discuss the example at some length as many of its properties are met with in other dicomplex systems.

If one adds a dilute HCl-solution to a warm mixture of gelatin and gum arabic (in water), one will see that the liquid becomes turbid at a pH-value below 4.8. With the aid of the microscope one can see that the turbidity must be ascribed to minute coacervate drops. This coacervation is readily reversible; addition of some NaOH makes the coacervate drops disappear. The reversibility is not restricted to the system gelatin/gum arabic, but holds generally, provided no secondary changes occur. The co-acervation of serum albumin and gum arabic may serve as an example. On acidification of this mixture typical coacervation is obtained. The coacervation is completely reversible only immediately after its production. If we wait too long the system will no longer clear completely as serum albumin is denatured in course of time.

The *rôle of pH* is very important in dicomplex coacervation. We have seen already that the coacervation between gelatin and gum arabic takes

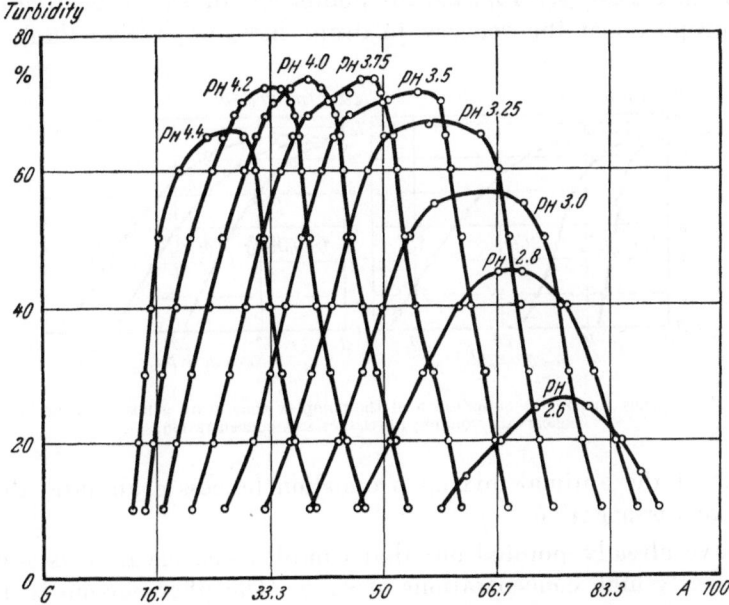

Fig. 25. Turbidity curves for mixtures of 0.05 % gelatin and gum arabic sols at different values of pH. Abscissae: mixing proportion expressed in per cent of the gum arabic sol.

place only at pH-values below 4.8. This represents the isoelectric point of the gelatin used. On the other hand the pH must not be too low, as in that case the carboxyl groups of the arabinate will get discharged. From this we see that this dicomplex coacervation takes place in a zone of pH where gelatin is positively [8] and gum arabic negatively charged.

It will be clear that at a certain pH the tendency to form a coacervate will be the strongest when the positive charge of the gelatin equals the negative charge of the gum arabic. Thus we will find an *optimum mixing proportion.* We can foresee that these optimum mixing proportions will depend on the pH-value. At pH = 2.6 e. g. only a very small amount of the carboxyl-groups of the gum arabic will be ionised (then the arabinate has, so to say, a large equivalent weight). At this pH-value we will need

---

[8] Of course, at pH below the isoelectric point the gelatin will still possess negative groups. As regards dicomplex formation we are only interested in the excess of charge in the protein molecule (the algebraic sum of its charges).

much gum arabic to compensate the positive charge of the gelatin. Fig. 25 shows some experiments on the influence of pH and mixing proportions on coacervation. At pH = 4.4 we need relatively much gelatin, at pH = 3.0 we find the reverse situation. Further we draw the attention to another phenomenon: the coacervate is soluble in excess colloid cation or colloid anion [9].

At the optimum mixing proportion the positive charge of the gelatin equals the negative charge of the arabinate. In agreement with this view, the coacervate drops show no electrophoretic mobility at this point. When gelatin is in excess (see Fig. 25; all points left of the respective optimum mixing proportions) the coacervate drops have a positive charge; at the

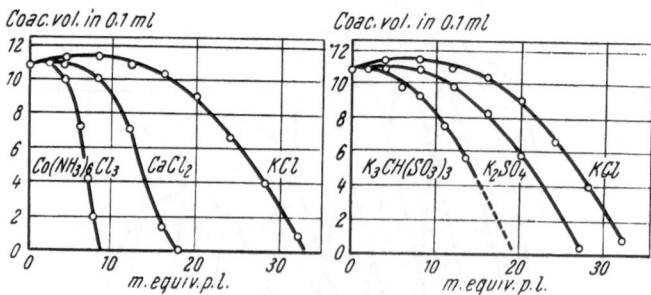

Fig. 26. Double valency rule in the suppression of the complex coacervate gelatin/gum arabic. Ordinates: coacervate volume; abscissae: salt concentration.

other side of the optimal mixing proportion (excess arabinate) the charge of the drops is negative.

We have already pointed out that complex coacervation is suppressed by sufficiently high concentrations of salts. The phenomenon is described by the *double valency rule*:

$$4\text{--}1 > 3\text{--}1 > 2\text{--}1 > 1\text{--}1,$$

$$1\text{--}4 > 1\text{--}3 > 1\text{--}2 > 1\text{--}1$$

(the suppressive action decreases to the right).

In illustration we give the results of some experiments in Fig. 26.

In biological systems, however, it will not often happen that the amounts of positive and negative charges are approximately equal, as most biocolloids are negatively charged. That is why an ideal double valency rule as depicted in Fig. 26 will not be found often. We will now discuss the kind of results one would get in non-ideal cases.

First of all let us see what influence the various salts have on the charge of the coacervate drops. We choose coacervate drops with a slight excess of gelatin and measure the electrophoretic mobility in the presence of salts

---

[9] In many serological flocculations (see page 40) the same phenomenon is observed.

(Fig. 27). The result is quite different from what we saw in Fig. 26; we find a *continuous valency rule*:

$$4–1 \ldots 3–1 \ldots 2–1 \ldots 1–1 \ldots 1–2 \ldots 1–3 \ldots 1–4$$

(relative positivation)   (relative negativation)

When working with a 3–1 salt the trivalent cation will have a strong affinity to the negative groups of the arabinate. Thus we have a relative positivation, while at the same time the complex relations are weakened.

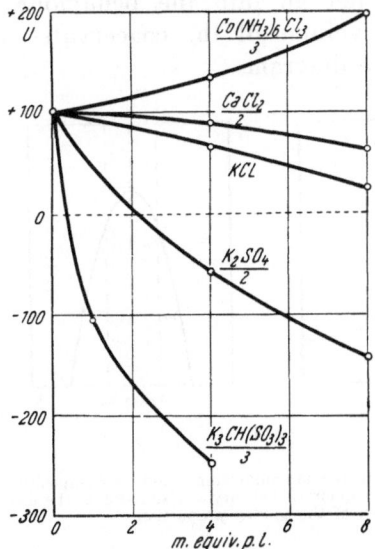

Fig. 27. Continuous valency rule in the influence of added salts on the electrophoretic velocity of complex coacervate drops (gelatin/gum arabic, with slight excess gelatin).

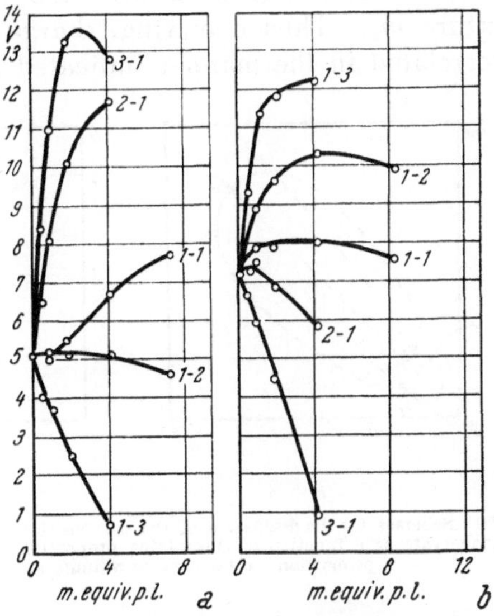

Fig. 28. Influence of small salt concentrations on the coacervate volume V of a strongly negatively charged (left) and a strongly positively charged coacervate (right). Gelatin/gum arabic coacervate.

In some cases the continuous valency rule can occur in experiments on the suppressive action of salts, viz. when the mixing proportion of the colloids is far from optimal (Fig. 28). When we take a complex with a large excess of arabinate, the complex relations will be rather weak. Addition of a 3–1 salt will result—as the affinity of the trivalent cation to the arabinate is strong—in a decrease of the negative charge. The proportion between positive and negative groups is improved and more coacervate will appear. Higher concentrations of the 3–1 salt will— as is expected from the foregoing considerations—have a suppressive action on the coacervate. This suppresion will again show the double valency rule. Thus we may get rather intricate diagrams in which the continuous valency rule as well as the double valency rule play a part.

The coacervate is in thermodynamical equilibrium with a colloid-poor liquid. To describe this equilibrium we may use the diagrams customary in the phase theory. We will not give a comprehensive discussion of this field, but only indicate the kind of diagrams obtained in this way

(Fig. 29). The diagram 29 a shows the composition of coacervates and equilibrium liquids. When mixing two solutions A and B, the one containing gelatin and the other arabinate, we pass a region where the mixture separates into two layers. The mixture m_1 e. g. forms the coacervate c_1 and the equilibrium liquid e_1. From the diagram one may read that coacervate and equilibrium liquid contain more arabinate than gelatin; consequently the volume of the coacervate is relatively small (Fig. 29 c) and it has a negative charge (Fig. 29 b). The mixture m_2 contains the optimal proportion of the colloids; the coacervate c_2 has the maximal volume (Fig. 29 b). We need not go into the behaviour of mixture m_3. Thus demixing, charge and volume of the coacervates are interrelated in the manner indicated in the diagrams.

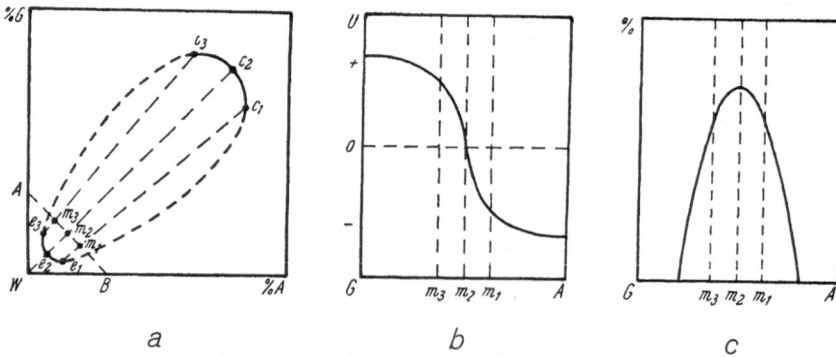

Fig. 29. Schemes for the discussion of the composition of coacervate and equilibrium liquid. *a* Composition of the coacervate as a function of the mixing proportion. *b* Electrophoretic velocity as a function of the mixing proportion. *c* Coacervate volume as a function of the mixing proportion.

It has already been stated that complex relations may play a rôle in other colloid systems too. The reason why so much work has been done on complex coacervates is simply that these systems show so many criteria by which the complex relations may be measured.

It has been shown by BUNGENBERG DE JONG and LANDSMEER that complex gels behave in a manner which can be predicted from the properties of complex coacervates. It is a well-known fact that gelatin gels swell considerably when placed in dilute acetic acid. A gel of a suitable mixture of gelatin and gum arabic does no longer swell in dilute acetic acid and moreover a considerable turbidity occurs in the gel. This is exactly what we would expect. The gelatin becomes positive and complex relations to the negative arabinate result in a suppression of the tendency to swell.

It has been possible to produce gelated complex coacervate drops of microscopic dimensions. The swelling of these drops may be measured easily under the microscope. As they have very small dimensions the swelling equilibria are established rapidly (e. g. within 5 minutes). On varying the pH of the medium the diameters of the gelated drops go through a minimum at a pH (3.7) which is approximately the same where (at a higher temperature) complex coacervation shows an optimum (at a given mixing proportion of gelatin and gum arabic).

Now the mixing proportion and the pH were kept constant and the influence of salts was investigated. The swelling of the drops (Fig. 30) is clearly governed by the double valency rule. So we see that the same factors which are determinative for complex coacervates also hold with these complex gels.

The complex gels mentioned, however, do not show a complete reversibility. In the original complex gel one of the partners is immobile (gelatin), while the other is in principle mobile, but the latter is bound to the gelatin gel structure by strong complex forces. When—by a change

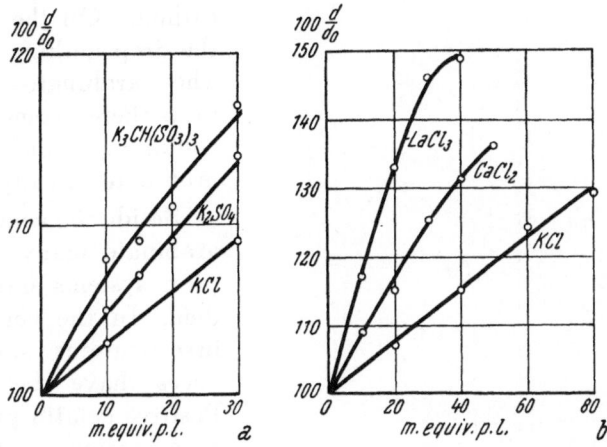

Fig. 30. Swelling of gelatinised coacervate drops (gelatin/gum arabic) by salts. Ordinate: diameter of the gel drops in per cent as compared with the original diameter. Abscissae: salt concentration.

of pH or by addition of salts—the complex relations are diminished, the mobile partner may be lost to the medium.

From the foregoing it will be clear that the binding between the two colloids in a complex system is not very strong. This can also be demonstrated by bringing the coacervate drops in an electric field (direct current). Three phenomena can be observed simultaneously: 1. electrophoresis, 2. deformation and 3. disintegration. The electrophoresis has already been discussed; the charge of the coacervate drop depends on the mixing proportion of gelatin and gum arabic and on pH. Deformation also occurs in an alternating current field. It consists of a flattening of the coacervate drops in a direction perpendicular to the lines of force of the field. The phenomenon seems to depend on the fact that the coacervate has a smaller conductivity than the equilibrium liquid. The disintegration phenomena (and electrophoresis, of course) only occur in a direct current field. A negative drop (Fig. 31), brought in an electric field, shows the following behaviour. Small vacuoles are formed within the drops. After some time these are united to large vacuoles, which are transported to the anode and then expel their contents outwards.

At the other side of the coacervate drop a "turbidity" appears, which on closer inspection is found to consist of numerous small coacervate drops.

With positively charged coacervate drops we see the mirror image of that with negative drops.

The explanation of these disintegration phenomena seems to be that the two colloid components in the complex coacervate are only loosely bound. Then in an electric field the gelatin cations will move to the cathode and the arabinate anions to the anode. In the case of a negative coacervate drop an excess of arabinate is present in the total system. The gelatin molecules—leaving the drop at the side of the cathode—find an excess of arabinate in the medium. Thus new coacervate drops will be formed at the side of the cathode. On the other side of the drop nothing will happen. The arabinate—leaving the drop there—comes in a liquid containing an excess of arabinate already.

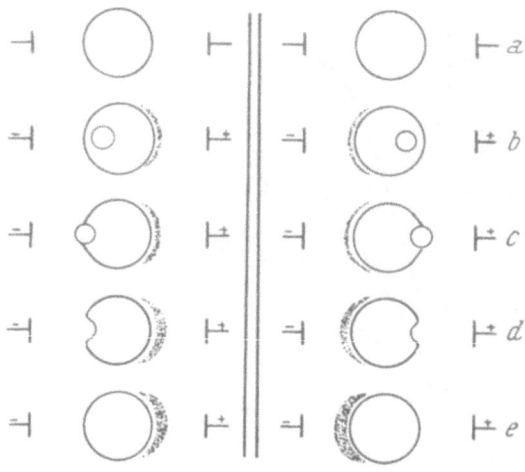

Coacervate drops positive   Coacervate drops negative

Fig. 31. Behaviour of complex coacervate drops in a direct current electric field (schematic).

Beside the complex gelatin/ arabinate many other dicomplex systems have been studied. In the course of these investigations specific differences have been observed. Positive gelatin gives complex coacervation with nucleate (and arabinate), with agar only a slight opalescence is produced, while complex coacervation (or flocculation) is entirely absent with soluble starch or glycogen. The NaCl resistance (= concentration of NaCl just sufficient for suppression of the complex relations) decreases strongly in the same direction. Thus we may arrange these colloids in a series of decreasing complex relations: nucleate > arabinate > agar > amylum solubile > glycogen.

We met this series in Table 1; it is the series of increasing equivalent weight. The complex relations are more intense the smaller the equivalent weight (= the greater the density of charge). Fig. 32 shows the difference between nucleate and arabinate. Two isohydric (pH = 3.7) series of mixtures were prepared—2% sols of gelatin (*G*) and nucleate (*N*) and 2% sols of gelatin and arabinate (*A*). The coacervate volumes and the reversal of charge points (arrows) have been measured. The results could be predicted from the fact that the equivalent weight of nucleate is much smaller than that of arabinate. For the combination *G* + *N* the maximal coacervate volume is less than that of the combination *G* + *A*. This relatively low water content is also to be expected, as the complex relations in the *G* + *N* combination are stronger (that is to say, more numerous per unit of weight).

We have seen that "ideal" phosphatides have no charge and that they may be considered as unicomplex systems. Natural phosphatides contain phosphatidic acid and consequently bear a negative charge. Thus, the phosphatide micelles may be seen as negatively charged units. Then we may expect interaction between these negative units and positive colloids. Indeed, complex flocculation is observed when mixing a gelatin sol and a phosphatide "sol" at low pH. Here too the complex relations depend on pH. At pH = 4.5 relatively large coacervate drops appear, but at pH = 3.5 only flocculi are produced. These flocculi consist of a very viscous coacervate and the water content is low. Addition of salt transforms the flocculi into liquid coacervate drops, while at still higher salt concentrations the coacervate is suppressed. Fig. 33 shows the result of

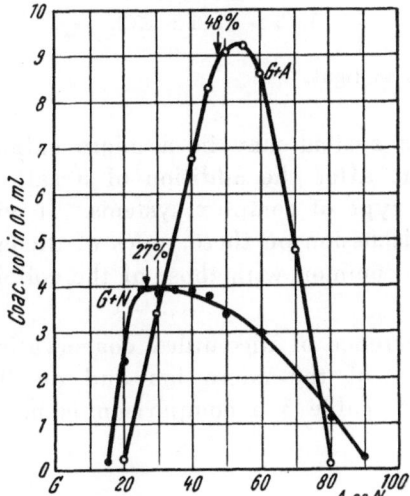

Fig. 32. Comparison of the coacervation complex of gelatin (G) + nucleate (N) and that of gelatin (G) + arabinate (A). Arrows indicate the reversal of charge points.

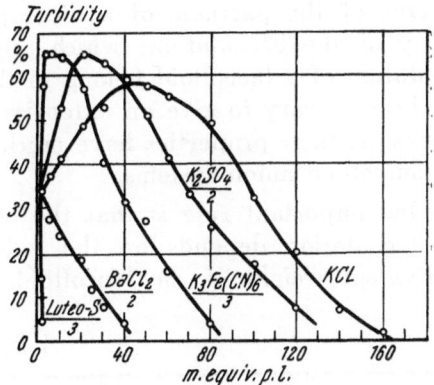

Fig. 33. Continuous valency rule and double valency rule in the action of salts on a positively charged complex coacervate (gelatin + soybean phosphatide).

a series of experiments on the influence of salts on a positive gelatin-phosphatide coacervate. In accordance with the fact that the mixing proportion of the colloids is far from optimal [10], we find firstly a continuous valency rule. As regards the suppressive action, however, the double valency rule is again obeyed.

In the preceding chapter we saw that the various biocolloids have specific "ion spectra" as regards the reversal of charge. From this we may expect that the suppression of complex relations by salts will also show specific ion sequences. A comparison of the following table with the sequence of the reversal of charge concentrations (see page 19) shows the agreement between the two sets of experiments. The sequence of the anions in Table 2 is the same in all cases. This is not surprising as the positive protein is a partner in all examples of the complex systems mentioned. The negative colloid partner has been varied and here we

[10] See page 31.

see—in the ion sequences of the cations—that the stronger the affinity to the negative groups of these biocolloids the stronger the suppressive action.

Table 2. *Ion sequences in suppression of complex relations.*

Positive protein + sulfate colloid and positive protein + carboxyl colloid	KCNS > KNO$_3$[1] KJ > KBr > KCl KCl > NaCl > LiCl
Positive protein + phosphate colloid	KCNS > KNO$_3$[1] KJ > KBr > KCl LiCl > NaCl > KCl

[1] The suppressive action decreases from left to right.

One of the partners of a dicomplex system may be a micro cation. Many of the flocculations which appear after the addition of a salt to a solution of a biocolloid belong to this type of complex systems. It will not be necessary to give an extensive discussion on these types of dicomplexes, as their properties have much in common with those of the colloid cation/colloid anion systems.

One important rule is that the occurrence of dicomplex coacervation or flocculation depends on the valency of the micro ion and on the equivalent weight of the biocolloid. In Table 3 a comparison is made

Table 3.

Colloid	Flocculating (or coacervating) salts	Non-flocculating or non-coacervating salts	Reciprocal hexol number
Na-pectate	6–1, 5–1, 4–1, 3–1, 2–1, 1–1		203
Na semen lini mucilage	6–1, 5–1, 4–1, 3–1	2–1, 1–1	563
Na-arabinate	6–1, 5–1, 4–1	3–1, 2–1, 1–1	1,068
Amylum solubile . .		6–1, 5–1, 4–1, 3–1, 2–1, 1–1	2,6000

between some biocolloids as regards the valency of the cations needed to cause flocculation or coacervation. The lower the equivalent weight the stronger the "combining power."

It will cause no wonder that these dicomplex systems too will be suppressed by the addition of salts. The resulting double valency rule, however, will be strongly asymmetrical in many cases. The suppression of a gum arabic/hexol nitrate [11] coacervate by salts, given in Fig. 34, shows this asymmetry in the double valency rule. A coacervate of a positive colloid (e. g. ichthyocoll) and a polyvalent anion (e. g. ferricyanide) gives

[11] See page 26.

the reverse phenomenon. Here too a double valency rule appears, but the cations are much more spread out than the anions.

Within a series of ions of the same valency typical sequences come to the fore. It is possible to obtain coacervation or flocculation of phosphatides with cations provided that certain conditions are fulfilled. The cation sequences obtained are (increase of concentration needed):

$$Ca < Mg < Sr < Ba \qquad \cdots\cdots\cdots \qquad Li < Na < K.$$

This is exactly what we would expect from the experiments on the reversal of charge of the phosphatides. In general one may say that with a given colloid the chance of dicomplex coacervation or flocculation (of the type colloid/ion) increases the lower the reversal of charge concentration of the ion in question.

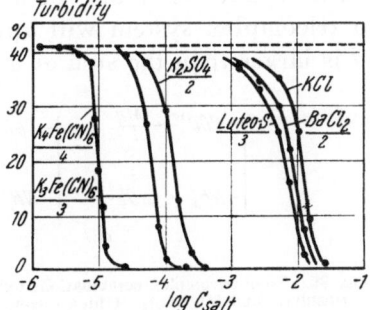

Fig. 34. Asymmetry in the double valency rule (gum arabic/hexolnitrate coacervate).

Finally we may mention that many of the dicomplex coacervations of this type may be supported by alcohol, acetone or other organic solvents. We have seen (Table 3) that gum arabic does not form a visible flocculation with salts of the types 3–1, 2–1 and 1–1. After the addition of acetone (relatively high concentrations are required) flocculation with these salts too can be produced.

Two factors might cause this phenomenon. Firstly the solubility of the colloid is in most cases decreased by the addition of the organic substance. Secondly it has been shown that the concentrations of cations needed for the reversal of charges of the colloid decrease considerably after the addition of alcohol or acetone. This means that the attachment of the cations to the negative groups of the colloids is strengthened considerably by the organic substances (lowering of the dielectric constant). Generally speaking, in macromolecular biocolloids both factors work in the same direction. In association colloids, however, the factors may have an opposite action. There the solubility is often increased by the addition of alcohol or acetone. In this case the flocculation or coacervation by a salt may be suppressed by an organic substance.

c) Tricomplex systems

A third type of complex system is characterised by the fact that three essential components must be present simultaneously. We will take an example:

$$isoelectric\ gelatin\ +\ Mn(NO_3)_2\ +\ K\text{-}chondroitinsulfate.$$

The number of theoretically possible interactions is rather large. The gelatin, being an amphion, might form a unicomplex system. The complex relations are too weak, however; the gelatin solution remains clear.

Three other combinations 1) gelatin $+ Mn(NO_3)_2$, 2) gelatin $+$ chondroitin sulfate and 3) $Mn(NO_3)_2 +$ chondroitin sulfate) do not show an interaction. Mixing the three components results in complex flocculation.

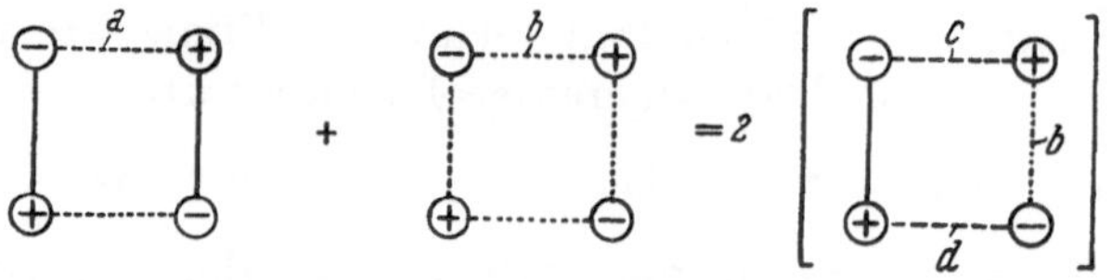

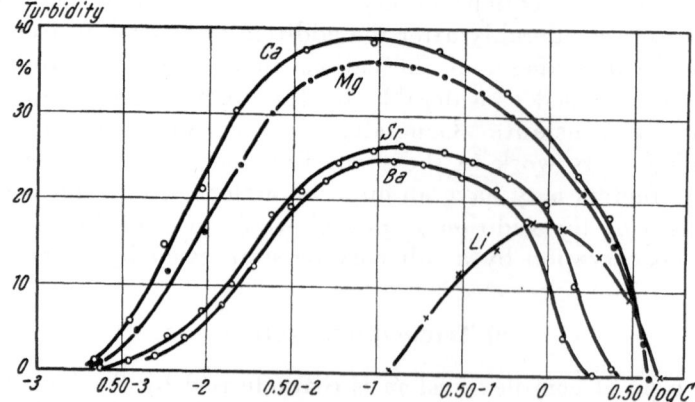

Fig. 35. Formation of a tricomplex system from a unicomplex and a dicomplex system.

One must always bear in mind that the components might be united in a scheme for a uni- or a dicomplex system (Fig. 35). It is clear that a tricomplex system will appear only when the sum of the forces c and d is larger than the sum of a and b.

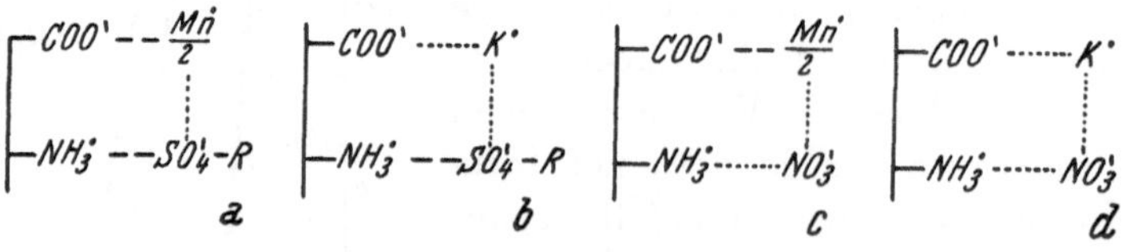

Fig. 36. Four tricomplex combinations exist in theory when dealing with a mixture of gelatin, K-chondroitinsulfate and $Mn(NO_3)_2$. (Intense complex relations: ⸺; negligible complex relations: ...).

In our example there are, theoretically speaking, four possibilities of tricomplex formation (Fig. 36). From the experiments described above it may be concluded that only the combination gelatin, chondroitin sulfate and Mn-ions leads to the production of a visible tricomplex.

Fig. 37. Tricomplex flocculation in a mixture of constant amounts of soybean phosphatide and carrageen and varying salt concentrations (abscissae: logarithms of the salt concentrations in eq./l.).

From the preceding sections it will be clear that here too specific ion sequences will be found. In the case of a colloid amphion and a colloid anion we will see a sequence of the micro cations which is specific for the negative groups of the amphions. The tricomplex flocculation of lecithin (colloid amphion), carrageen (colloid anion) and various cations gives a clear example of this rule (Fig. 37). The tricomplexes show the sequence: Ca > Mg > Sr > Ba > Li (decreasing action of the cations).

Compare the ion spectra of phosphatides in Fig. 14. We also draw attention to the fact that an excess of salt makes the tricomplex disappear.

It will cause no wonder that tricomplex flocculation can also be obtained in the combination colloid amphion/micro cation/micro anion. This interaction too may be represented by the scheme of Fig. 35. Of course, the micro ions must have a rather strong affinity to the charged groups of the colloid. A low affinity of the anion may, however, be compensated by a strong affinity of the cation. The reverse of this rule also holds. Thus one can produce flocculation of isoelectric gelatin with a combination of K_2HgJ_4 (strong affinity to positive groups) and KCl (low affinity to negative groups). The combination of KCNS and KCl does not give tricomplex flocculation; here it is necessary to replace the K-ions by e. g. Cd-ions (which have a much stronger affinity to negative groups).

d) Some biological examples

1. Clotting of blood

Many authors think that the second phase of blood coagulation is an enzymatic process, thrombin acting in some unknown way on fibrinogen. VAN DER MEER (1946) investigated the binding between thrombin and

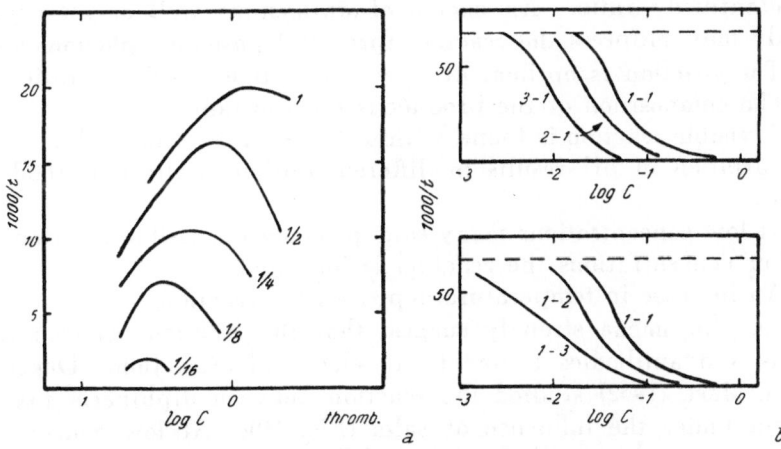

Fig. 38. *a* Velocity of clotting when keeping constant the fibrinogen concentration, while the thrombin concentration is varied (1, $^1/_2$, $^1/_4$, etc. = dilution of fibrinogen).
b Influence of salts on the clotting process.

fibrinogen. He found several interesting phenomena (Fig. 38). First of all it must be noted that the velocity of clotting shows a maximum when keeping the fibrinogen concentration constant and varying the thrombin concentration (Fig. 38 a). Compare the dicomplex systems: the complex relations are strongest at a certain mixing proportion (Fig. 25). Furthermore the amount of thrombin needed for optimal clotting diminished as the fibrinogen concentration is decreased. The influence of salts too suggests that complex relations play an important part (Fig. 38 b). The double valency rule comes to the fore when an excess of thrombin is present. At the optimal mixing proportion (smallest clotting time) the salts show an influence comparable to Fig. 28 (compare Fig. 26). Moreover the resistance to the influence of neutral salts is largest at the optimal proportion of the colloid components. On the basis of these experiments Van der Meer comes to the hypothesis that complex formation plays an important rôle in the second phase of blood coagulation, thrombin providing the positive charges, fibrinogen the negative ones. In this respect it is important that quinine may produce a (reversible) clot when added to a fibrinogen solution. After the binding of thrombin to fibrinogen by complex forces the first component may exert an action on the second one, resulting in the irreversibility of the clot. Perhaps this will be an enzymatic action, though it must be remembered that in the complex coacervate gum arabic/serum albumin the protein is slowly denatured. Thus it is possible that the complex binding alone suffices for a "denaturation" of fibrinogen.

2. The reaction between antigen and antibody

The following phenomena have been found in many serological reactions:

a) A visible reaction appears only at rather narrow limits of the antigen/antibody ratio. An excess of antigen as well as an excess of antibody may suppress the reaction (pre- and post-zone phenomenon).

b) The reaction is optimal at a certain antigen/antibody ratio.

c) The composition of the product is not constant.

d) A visible reaction is found within definite pH-values only.

e) Variation in pH results in different optimal antigen/antibody proportions.

f) At low concentrations many salts promote the visible reaction, while at higher concentrations the reaction is inhibited.

g) An increase in temperature depresses the reaction.

These phenomena strongly suggest that the reaction between (many) antigens and antibodies is due to an electrical attraction. Dekker and Van der Meer (1952) studied the reaction between diphtheria toxin and antitoxin under the influence of salts (Fig. 39). At low concentrations the continuous valency rule appears, while the double valency rule is seen at high concentration. Their conclusion is that we are dealing with an unbalanced complex system (excess of negative charges). It is not certain whether a dicomplex or a tricomplex is involved.

3. Connective tissue

The fact that in connective tissue an amphion (collagen) and a colloid anion (mucopolysaccharide) are present leads to the question whether these components are linked together by complex forces. LOEVEN (1953) studied the swelling of connective tissues under the influence of pH and of added salts. Some of his results have been summarised in Fig. 40. The influence of pH depends very much on the amount of muco-polysaccharide found in the tissue. Moreover the type of curve found for cartilage may be changed via the cornea-type to the sclera-type by gradual extraction of chondroitin-sulfate from the tissue. Model systems (gels of pigskin-gelatin with an admixture of chondroitin sulfate) behave in the same manner. At the minimum of swelling (maximal complex forces) the influence of salts has been found to follow the double valency rule.

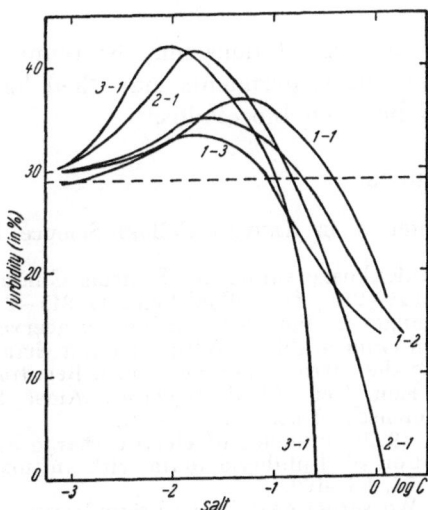

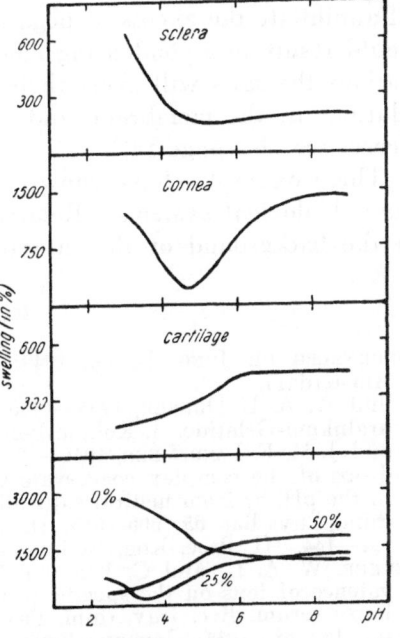

Fig. 39. Effect of salts on the precipitation re-action (diphtheria toxin/antitoxin).

Fig. 40. Influence of pH on the swelling of connective tissue as compared with models (containing pigskin gelatin and K-chrondroitin-sulfate). Per cent chondroitinsulfate indicated in the lower diagram.

At this point (low pH) a dicomplex will be present. It should be noted, however, that this dicomplex cannot exist at physiological values of pH. Here we must suppose that the linkage between the two colloids is of the tricomplex type. Especially the concentration of the calcium ions is of great importance for the integrity of the system. The mono-valent ions do not play an important rôle. This is exactly what we would expect in the case of a tricomplex system.

4. The protoplasmic membrane

Some twenty years ago BUNGENBERG DE JONG suggested that the proto-plasmic membrane consists of a complex system of phosphatides. DE HAAN (1935) performed some experiments on the permeability of water in *Allium* cells and found that this hypothesis was in agreement to the properties of the cell. He measured the permeability of water (plasmolytic method) under the influence of salts with various cations (3–1, 2–1 and 1–1 salts, viz. $Co(NH_3)_6Cl_3$, $Ca(NO_3)_2$ and $NaNO_3$) which were added to the plasmolyticum (sugar). Small concentrations of a 3–1 or 2–1 salt decrease the permeability of water. At higher concentrations all salts increase the permeability. As regards the decreasing influence one obtains the series:

$$3\text{–}1 > 2\text{–}1 \; (> 1\text{–}1).$$

This tallies with the hypothesis that di- and trivalent cations will first of all annihilate the excess of negative charge in the phosphatide layer. This would result in a condensing effect on the membrane. At higher concentrations the salts will exert their normal decreasing effect on the complex relations in the membrane and then the permeability for water will increase (see also page 147).

These examples have shown that complex relations may be found in many biological systems. Relatively simple experiments may shed light on the background of the interactions between biocolloids.

References

Bungenberg de Jong, H. G., 1949: Chapter X in Kruyt's Colloid Science II, Amsterdam.
— und W. A. L. Dekker, 1935—1936: Complexkoazervation des Systems Gummiarabikum-Gelatine. I. Kolloid-Beih. 43, 143—212; II. Kolloid-Beih. 43, 213—271.
— and J. M. F. Landsmeer, 1946—1947: Changes in diameter of gelated coacervate drops of the complex coacervate Gelatin-Gum arabic, resulting from a change in the pH, or from neutral salts added to the surrounding medium. I. Rec. trav. chim. Pays-Bas 65, 606—613, II. Proc. Kon. Ned. Akad. Wetensch. Amst. 51, 137—144, III. Proc. Kon. Ned. Akad. Wetensch. Amst. 51, 295—301.
Dekker, W. A. L., and C. Van der Meer, 1952: Influence of electric charge and valence of ions on the specific precipitation of diphtheria toxin with antitoxic horse serum. Rec. trav. chim. Pays-Bas 71, 88—100.
Haan, Iz. de 1935: Ionenwirkung und Wasserpermeabilität. Protoplasma 24, 186—197.
Loeven, W. A., 1953: Het complex systeem collageen-mucopolysaccharide in bindweefsel. Thesis, Leiden.
— 1955: The binding collagen-mucopolysaccharide in connective tissue. Acta Anat. 24, 217—244.
— 1955: The nature of the complex binding between collagen and mucopolysaccharide in connective tissue. Acta Physiol. Pharmacol. Neerl. 4, 243—273.
— 1956: The complex binding between protein and mucopolysaccharide in connective tissue. Acta Physiol. Pharmacol. Neerl., in press.
Van der Meer, C. 1946: Complex relations in the 2nd phase of blood coagulation. Proc. Kon. Ned. Akad. Wetensch. Amst. 49, 251—264.
— 1948: Complexbetrekkingen in de tweede phase van de bloedstolling. Thesis, Leiden.

5. Association Colloids

a) Introduction; micelle types

The association colloids have been defined as follows. The smallest kinetic units (molecules or ions) do not have colloidal dimensions. The associations, in thermodynamical equilibrium with the small units, fall into the class of the colloids.

The association colloids comprise a. o. soaps and synthetic detergents with a negative group (e. g. sulfate-soaps) or a positive group (e. g. the quaternary ammonium group). These association colloids do not take part in the building of protoplasm. Yet physical biochemistry is highly interested in these substances, as phosphatides play an important rôle in protoplasm. Phosphatides too should be reckoned among the association colloids. A rational way to the study of phosphatides is possible only after some insight has been gained as to the structure of the associations of the relatively simple detergents.

Normally the association of the detergents is studied in the system soap/water. This type of study led to the result that every soap has a

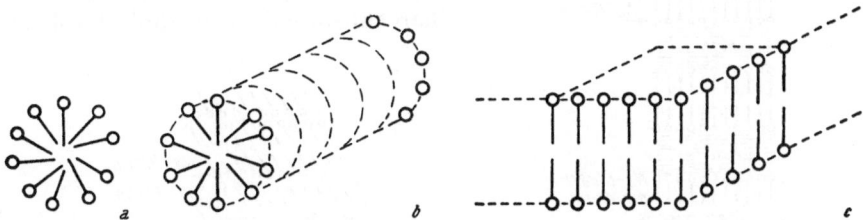

Fig. 41. Types of micelles in soap systems.

characteristic critical concentration of micelle formation. This critical concentration depends strongly on the length of the carbon chain. At higher concentrations the soap is present in the form of micelles beside the free molecules (or ions). It has been suggested that at concentrations just above the critical one we will find spherical micelles. Then a number of soap ions form an aggregate of disorderly—perhaps even folded—carbon chains, while the ionized groups are situated at the periphery of the more or less spherical micelle (Fig. 41 a).

X-ray investigations of concentrated soap solutions showed a more regular structure of the soap molecules. Now the soap molecules are united in double layers, which layers in their turn are the components of a secondary association (Fig. 42). At sufficiently low temperature some soaps will crystallise from their solutions in water (e. g. Na-stearate at room temperature). The crystals show—according to X-ray studies—double layers of soap molecules, but here the spacing found is smaller than the double length of the molecule (Fig. 43). The molecules are placed at a definite angle to the plane of the polar groups.

One might ask which of these three structural elements (spherical micelle, sandwich micelle—double layer with ordering in two dimensions—, and crystal—ordering in three dimensions—)plays the most important part in biology. We suppose that in the case of phosphatides comparable associates may be possible.

We think that phosphatides form an important component of interfacial membranes, while on the other hand three dimensional structures—in combination with protein—are possible. The spherical micelle (Fig. 41 a) is not fit for these functions. Double layers may form three dimensional structures by accumulation, while they are the prototype of membranes. Presumably a membrane consisting of a crystalline double layer (Fig. 43) will be less permeable than a membrane of the type of a sandwich micelle (Fig. 41 c). Yet we do not think that the crystalline double layer plays a rôle in active cell life. Crystalline structures do not show variation as to their degree of packing. The living cell needs membranes, which may be regulated as regards their permeability. It seems that only the micelle in which the lipid molecules are placed perpendicular on the plane of the polar groups meets the necessities of a biological membrane. There we have the possibility of increase or decrease of permeability. Moreover, other substances, e. g. cholesterol, may

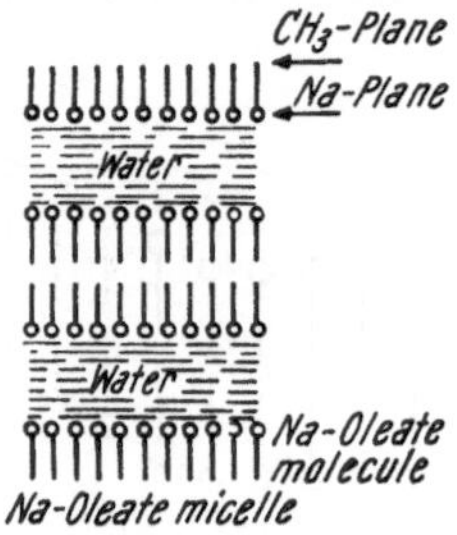

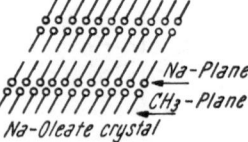

Fig. 42. Structure of the Na-oleate micelle (Hess, Kiessig und Philippoff 1941).

Fig. 43. Structure of the Na-oleate crystal.

take part in these membranes. In this case, in contra-distinction to the crystals, a mixture of phosphatides may build up the membrane.

Before entering the field of the biologically important association colloids (phosphatides) it will be necessary to study the simpler detergents. We will first give a description of the colloid systems containing micelles of the non-crystalline type (we will use the term sandwich micelle for all types of double layers where the lipid molecules are placed perpendicular to the plane of the polar groups). This type of micelle is present in 1. elastic-viscous systems, 2. coacervates and 3. the smectic phase (e. g. myelin tubes, which constitute a growth form of the smectic phase). As the latter systems have not been studied systematically we will pay attention practically exclusively to the first two systems.

b) Colloid systems of simple detergents, characterised by the presence of sandwich micelles

In a solution of Na-oleate we find a drastic change in micelle type when the Na-oleate concentration is increased (spherical micelles → secundary association of sandwich micelles—Fig. 42). It seems certain that the background of this phenomenon must be thought not only in the increase of the concentration of "micellar" material, but in an inactivation of the ionised groups at the surface of the micelles. In our laboratory we have chosen a way in which these two possibilities may be studied separately. In one case we study the effect of increasing salt concentration at a constant—relatively low—concentration of soap. In the second case a constant salt concentration is maintained, while the soap concentration is increased.

It has been observed that the first way leads to a number of radical changes in the soap system, while the second way shows only qualitative changes. Thus it is clear that the increasing inactivation of the ionised groups by added ions, is by far the most important factor for the changes in micelle type.

We take e. g. a 1% K-oleate solution and add increasing amounts of KCl. The viscosity of the original solution has about the same value as that of water, but it increases gradually at increasing KCl concentrations. At a certain KCl concentration ($\pm\,0.45\,n$) a steep increase is observed. Moreover, at this point a clearly visible elasticity comes to the fore, while the soap solution can also be spun into threads. We will call this KCl-concentration the *"elastic limit"* henceforth. A further increase of the KCl-concentration results in an increase of viscosity up to a certain maximum. The elastic properties too become ever more distinct, to disappear at the next characteristic KCl-concentration: the *"coacervation limit."* There the liquid separates into two layers: 1. a layer which contains practically the whole amount of oleate (the *oleate coacervate*) and 2. an oleate solution of very low concentration (the *equilibrium liquid*). The oleate systems in between the elastic limit and the coacervate limit unite in themselves the properties of liquids (they may be poured out) and solids (elasticity). Consequently they will be referred to as *elastic-viscous* systems. Chapter 6 will be devoted to these systems. Here we will only mention that we are dealing with completely clear systems, which from the point of view of the phase theory should be regarded as homogeneous solutions. Their peculiar properties must be ascribed to a strong association. Indeed, the elastic limit is not a clear cut boundary, where the elasticity suddenly appears; it is a practical boundary where the elastic properties become so strong as to be easily detectable. Besides it may be mentioned that elastic-viscous systems are isotropic when in rest, but that they show a strong double refraction of flow.

Starting at the coacervation limit we will now increase the KCl-concentration still more. It is typical for this system that the coacervate layer at a concentration near to the coacervation limit has a volume only slightly less than the total volume and that this coacervate volume decreases sharply when increasing the KCl-concentration. Finally one passes two more characteristic KCl-concentrations. In the first a smectic liquid phase begins to appear, which phase grows at the expense of the coacervate. In the second the whole coacervate layer has been transformed into a smectic phase. This phase is spontaneously birefringent. Fig. 44 gives a survey of the phenomena described.

When we study the influence of KCl on oleate solutions of other concentrations (e. g. 0.5, 1.5 or 2%), we find the same marked changes in the state of the systems. A change in the soap concentration, on the other hand, only results in quantitative differences. At higher soap concentrations, e. g. the rise in viscosity in the non-elastic range is larger, the maximum of viscosity in the viscous-elastic range is higher, the elastic properties are more pronounced, while the minimal coacervate volume finally reached is larger.

Summarising it may be said that an increase of the KCl concentration leads to drastic changes in the soap system, while an increase of the oleate concentration only results in quantitative changes. This points in the direction of the view that the various changes (non-elastic → elastic-viscous → coacervate) are closely connected with an inactivation of the ionised groups at the surface of the soap micelles. Up till now we have

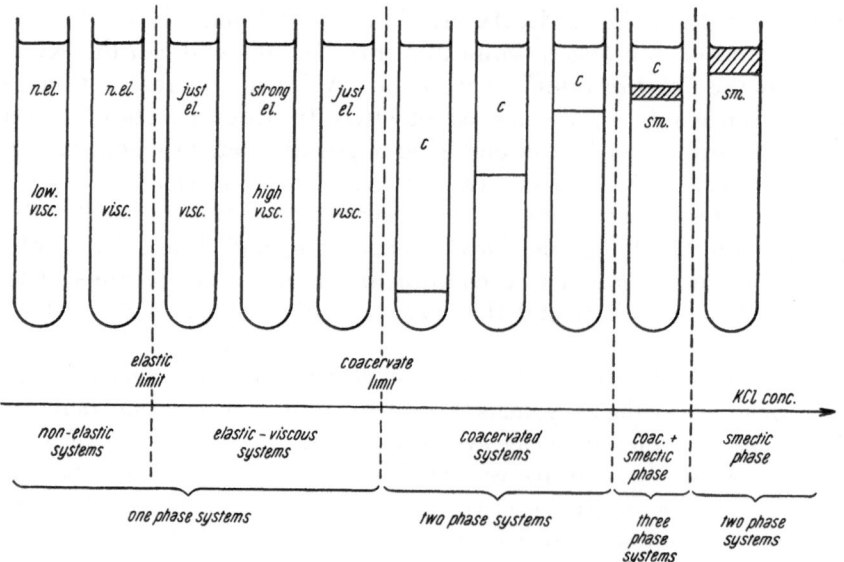

Fig. 44. Changes in an oleate solution produced by addition of KCl.

used the term "inactivation" in a wide sense. Two possibilities of "inactivation" are: a) compression of a diffuse double layer and b) binding of K-ions to the micelle surface (then we are dealing with a mixed micelle consisting of negatively charged and uncharged oleate).

One of us (H. L. B., 1949) framed the hypothesis that the formation of elastic-viscous systems and coacervates by salt is caused by an "inactivation" of the negative groups of the second type. He started from the consideration that in an associate (such as K-oleate) two factors play a rôle, which work in opposite directions (Fig. 45).

a) The carbon chains attract each other (London-van der Waals forces) and these forces would lead to a parallel ordering of the molecules.

b) The ionised groups repel each other; they try to take positions as far away from each other as possible. In the spherical micelle this situation is reached.

In a diluted soap solution without salt factor *b* will be the stronger of the two, thus here we expect only spherical micelles. The hypothesis supposes that added salts have a reducing influence on factor b, which results in changes in the association as regards the shape of the association as well as the number of soap molecules partaking of one micelle. The decharging effect of the salt results in the creation of uncharged groups at the surface of the micelle. The following changes are supposed to take place at increase of the salt concentration:

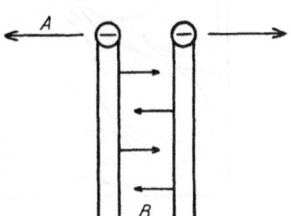

1. The spherical micelles (HARTLEY-micelles) are transformed into small sandwich micelles (Fig. 46 $a \rightarrow b$).

2. These small sandwich micelles grow out (at the expense of their number) into big flat micelles (Fig. 46 $b \rightarrow c$).

Fig. 45. Two opposing factors determine the structure of the soap micelles: A the repulsive Coulomb-forces of the charged groups and B the attractive London-Van der Waals forces between the carbon chains.

3. The large micelles may adhere locally by virtue of the undissociated groups at their surfaces. Then a three-dimensional net-work throughout the whole system comes into being. This supposition makes the high viscosity and the elastic properties understandable (Fig. 47 a).

4. At still higher salt concentrations the number of undissociated groups (and consequently the number of points of contact) increases steadily. The three-dimensional net-work withdraws from the medium (coacervation, Fig. 47 b).

This type of coacervation is characterised by the fact that it starts with a coacervate volume of 100%, which decreases at higher salt concentrations.

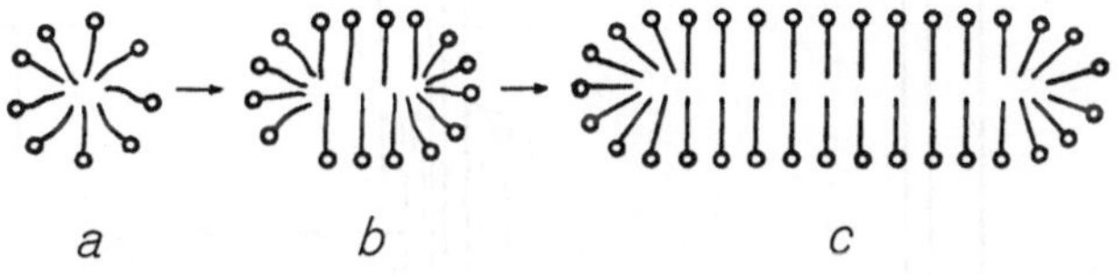

Fig. 46. Gradual transformation of small Hartley-micelles into large flat sandwich micelles.

5. The smectic phase, which originates at very high salt concentrations, agrees very well with the hypothesis. Here the sandwich micelles have joined together in such a degree that they are situated parallel in macroscopic dimensions. The system is now spontaneously birefringent (Fig. 47 c). Under the same circumstances oleate may form myelin tubes. Here we find—based on optical experiments—that the soap molecules are indeed situated parallel to each other and perpendicular to the surface of the myelin tube [12] (compare Fig. 5).

---

[12] A general method to make myelin tubes from any detergent has been worked out at our laboratory. In all cases studied the detergent molecules are oriented in the same manner as phosphatide molecules in myelin tubes.

The postulated presence of sandwich micelles in elastic-viscous oleate systems and oleate coacervates has been corroborated by experiments on the double refraction of flow by van den Berg (1953). Here we will first

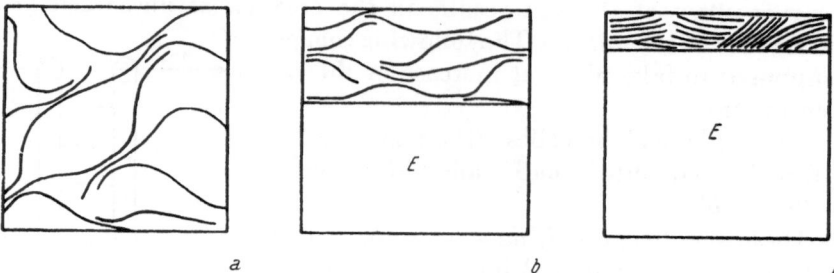

Fig. 47. Supposed interaction of the soap micelles in *a* the elastic-viscous system, *b* the coacervate and *c* the smectic phase (*E* = equilibrium liquid).

of all mention his qualitative observations on the position of the index ellipse relative to the orientation of the micelles. These experiments have been performed with the aid of three apparatuses, which allow for observations in three directions, perpendicularly to each other (Fig. 48).

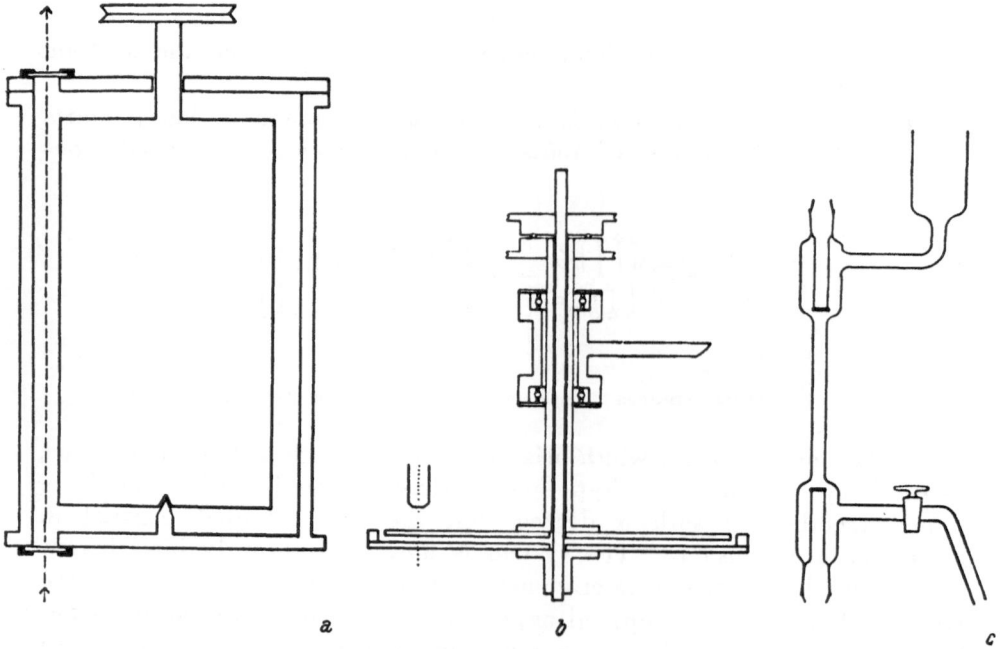

Fig. 48. Three apparatuses for determination of birefringence of flow in directions perpendicular to each other.

One may now—as the position of the carbon chains in the micelles is given by the long axis of the index ellipse—predict how the position of the index ellipse in the apparatuses A, B and C will be, when starting from three possible types of micelles (Fig. 49, *I, II* and *III*) which are oriented by the flow of the system.

I. The micelles consist of a pile of rather small sandwich micelles (compare Fig. 42).

II. The micelles are large sandwich micelles (compare Fig. 41 *c*).

III. The micelles have the shape of cylinders (compare Fig. 41 *b*).

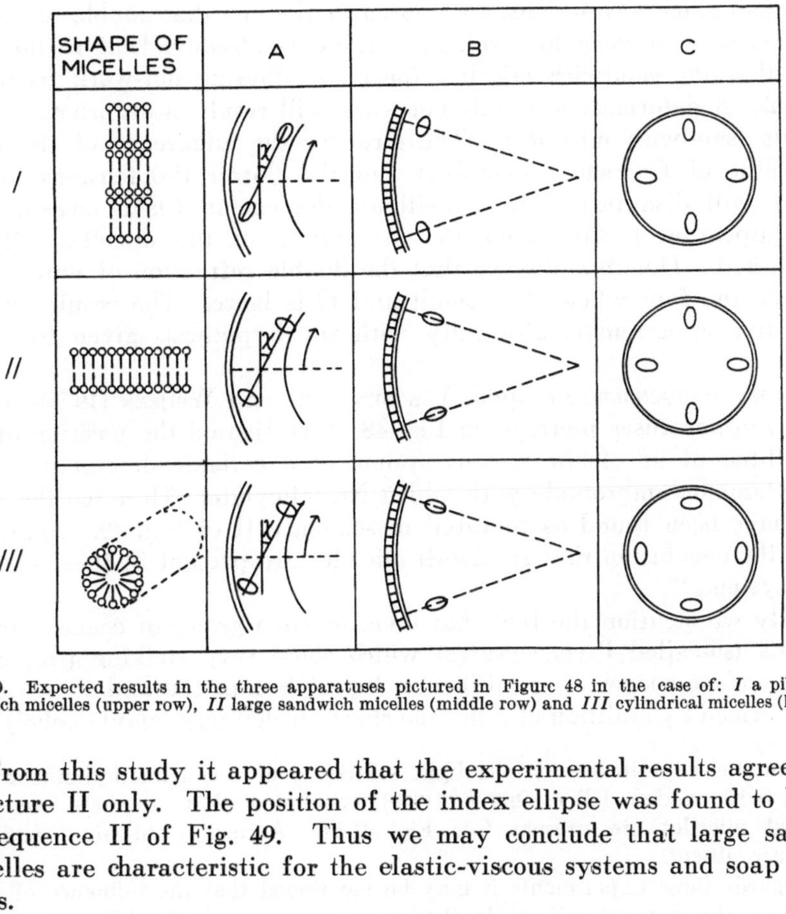

Fig. 49. Expected results in the three apparatuses pictured in Figure 48 in the case of: *I* a pile of small sandwich micelles (upper row), *II* large sandwich micelles (middle row) and *III* cylindrical micelles (lower row).

From this study it appeared that the experimental results agreed with structure II only. The position of the index ellipse was found to be that of sequence II of Fig. 49. Thus we may conclude that large sandwich micelles are characteristic for the elastic-viscous systems and soap coacervates.

The quantitative part of this study was restricted to an investigation of the double refraction of flow of elastic-viscous systems in the Kundt-apparatus (Fig. 48 a). Double refraction comes to the fore already at very low velocity gradients. That is why the apparatus has been constructed in such a way as to make measurements at a wide variety of velocity gradients possible. It was shown by VAN DEN BERG that a strong correlation exists between the double refraction of flow and the elastic properties. Both depend on the same variables (salt concentration, oleate concentration, temperature, addition of organic non-electrolytes, etc.). A correlation exists e. g. at constant oleate concentration and temperature and increasing KCl concentration which may be expressed in the following function:

$$\frac{1}{\gamma} = \frac{kG}{\varDelta}$$

Here $\gamma =$ velocity gradient in sec^{-1} where a certain extinction angle is reached; $G =$ shear modulus in dyne/cm^2; $\varDelta =$ logarithmic decrement (measured in spherical vessels of constant diameter) and $k =$ a constant.

This and other correlations, and moreover the fact that double refraction of flow is seen at very low velocity gradients already, lead to the conclusion that the sandwich micelles form an adhering net-work (compare Fig. 47 a). A deformation of this net-work will result in an orientation of the units (sandwich micelles). The stronger the adherence of the units (large value of $G =$ shear modulus) and the slower the tensions in the net-work will disappear (the logarithmic decrement $\varDelta$ is a measure for this disappearance), the easier the orientation of the micelles will be accomplished. Thus we observe that the double refraction of flow comes quickly to the fore when $\varDelta$ is small and G is large. The results of the quantitative experiments also tally with the hypothesis given by Booıj (1949).

Recently Bungenberg de Jong, Van den Berg and Weijzen (1955)—using the three apparatuses pictured in Fig. 48—investigated the position of the index ellipse of an elastic-viscous system of a cationic detergent (cetyl-trimethylammoniumbromide with added Na-salicylate). Here too the same results have been found as pictured in sequence II of Fig. 49. Again we come to the conclusion that sandwich micelles are present in these elastic-viscous systems [13].

Finally we mention the fact that there exists a group of coacervates of detergents (so-called P-coacervates) which show very striking properties at the interface coacervate/equilibrium liquid (see p. 130 and 137). They may be formed by addition of a not too short alcohol (e. g. amylalcohol) to a

---

[13] From his experiments on the light scattering of cetyltrimethylammonium-bromide (with added KBr) Debye (1949) concluded that in these solutions cylindrical micelles are present (see Fig. 41 b). According to him sandwich micelles are absent.

As regards these experiments it may be mentioned that the influence of salts on cetyltrimethylammoniumbromide shows very strong specific differences. The concentrations needed to give one of the successive changes (viscous → elastic-viscous → coacervate) are strongly dependent on the anion of the added salt. The Br-ion belongs to the group of ions of very low affinity. Even a saturated solution of KBr is not able to coacervate a cetyltrimethylammoniumbromide solution. Only a faint elasticity begins to appear at this very high concentration. The KBr-concentrations used by Debye are so low, that the solution has not yet reached the elastic limit. Only an increase of viscosity takes place at these concentrations. It is quite conceivable that this increase in viscosity must be ascribed to the formation of cylindrical micelles. The idea that the cylindrical micelle is an intermediate stage between the spherical micelle and the sandwich micelle would, moreover, fit very well in our hypothesis (compare Fig. 41).

suspension of the smectic phase (solution of detergent $+$ a large amount of salt). This type of coavervate interests us because it also appears in the case of phosphatides (BUNGENBERG DE JONG and WESTERKAMP 1932). There they may be realised by addition of organic polar substances which may occur in biological systems (oleic acid, triolein etc.). Alcohols also give the same phenomenon.

VAN DEN BERG examined one example of these oleate coacervates as regards the double refraction of flow. His results indicated that in this case too we are dealing with sandwich micelles.

It is convenient to distinguish the two types of coacervates of association colloids by separate designation. Those which arise from detergents by the action of salts alone and which show no trace of the striking properties at the coacervate boundary will be designed with *O-coacervates* (*O* stands for ordinary). The other type with striking surface properties will be called *P-coacervates* (*P* means phosphatide-like, because they were first met with in the study of phosphatides; BUNGENBERG DE JONG and DE HEER 1949; BUNGENBERG DE JONG and WEIJZEN 1951; BUNGENBERG DE JONG and DE BAKKER 1955). At present O-coacervates are also known in phosphatides (BUNGENBERG DE JONG, DE BAKKER, and ANDRIESSE 1955), so that it is better to forget the historical meaning of the prefixes.

The survey below gives a summary of the different types of colloid systems of simple detergents and the arrows give the ways in which they can be reached starting from the non-elastic solution.

Spherical micelles	Sandwich micelles			
non-elastic solution	elastic-viscous solution	O-coacervate	Smectic phase	P-coacervate
			$\longrightarrow$ - - - - - - - - -	$\longrightarrow$
		$\longrightarrow$ - - - - - -	- - - - -	$\longrightarrow$
	$\longrightarrow$ - - - - -	- - - - - - - -	- - - - -	$\longrightarrow$

The solid arrows mean increasing concentrations of a suitable salt (e. g. KCl in oleate) and the dotted arrows mean increasing concentrations of a suitable higher alcohol (e. g. amylalcohol).

The survey shows that with salt alone we may realise elastic-viscous systems, O-coacervates and the smectic phase, but not P-coacervates. The latter cannot be realised by the action of amylalcohol alone (or another suitable polar organic non-electrotrolyte), but only by the joint action of salt and amylalcohol.

The combinations of arrows in the survey show that P-coacervates become possible with amylalcohol when the salt concentration is sufficient to arrive at a colloid system which already contains sandwich micelles.

Thus for instance we may add so much salt that we attain to the elastic-viscous system and then addition of amylalcohol may ultimately lead to P-coacervates, whereby we pass successively the O-coacervate and the smectic phase.

The P-coacervates contain still many unsolved mysteries and in the following chapters they will be mentioned only a few times. When in the following coacervates of detergents are mentioned without more, we will always mean O-coacervates.

c) The binding of the oppositely charged ion
(The sandwich micelle regarded as a complex system)

The starting-point of our theory on the transformation of spherical micelles into sandwich micelles as sketched in the preceding section has been the hypothesis that the addition of salt caused the formation of non-ionised groups at the surface of the micelle. At the time of the framing of this hypothesis only examples of elastic systems and coacervates were known in which the concentrations of salt needed are relatively high. Under these circumstances it is very difficult to prove that some of the ions needed are really bound to the soap ions.

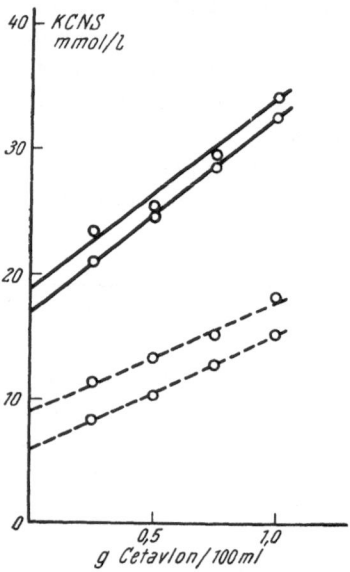

Fig. 50. KCNS concentrations needed to reach certain criteria in a CTAB (cetavlon) system. Upper straight line: concentration required for a coacervate volume of $C \times 24\%$; lower straight line: same for a coacervate volume of $C \times 40\%$; upper dotted line: same for the dynamic elastic limit; lower dotted line: same for the static elastic limit.

Later on Bungenberg de Jong and Recourt (1952) found a case in which the possibilities are much better as there the salt concentrations needed are exceptionally low. They studied the combination cetyltrimethylammoniumbromide [14] and KCNS (or KJ). Fig. 50 shows the coacervation limit and the elastic limit for the combination CTAB + KCNS as a function of the CTAB-concentration. It appears that the concentrations of KCNS needed to reach the coacervation limit or the elastic limit are linear functions of the CTAB concentration. Such linear functions would be expected from the supposition that a certain percentage of the ionised groups must be occupied in order to reach the elastic limit, respectively the coacervation limit. The slope of the lines is then a measure of the degree of occupation. From the CTAB concentration it may be computed that the coacervation limit is reached at a degree of occupation of 70% of the positive groups. For the elastic limit a value of 40% has been found. The points of intersection with the ordinate have the following meaning: they denote which equilibrium

---

[14] Which will be called henceforth CTAB.

concentration of CNS⁻-ions in the medium belongs to the degrees of occupation in question. This equilibrium concentration is larger for the coacervation limit than for the elastic limit. Thus Booɪɪ's hypothesis concerning the progressive binding of the oppositely charged ion at increasing salt concentration has been confirmed. In this combination we get elastic-viscous systems when the degree of occupation is higher than approximately 40% and coacervation when a value of 70% is reached.

First of all the influence of temperature has been investigated. If we have a mixture of CTAB of KCNS which just coacervates at a given temperature, we see that an increase of temperature causes the coacervate

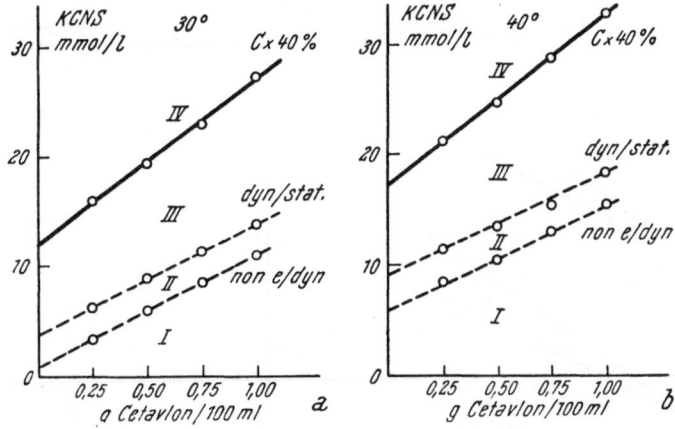

Fig. 51. KCNS concentrations required to reach certain criteria in a CTAB system at different temperatures.

to disappear. The coacervated system changes into an elastic system. The elastic limit is influenced in the same manner. When the KCNS concentration is high enough to give a barely visible elasticity, we can make the elasticity disappear by increasing the temperature.

The background of this phenomenon may be found by an investigation at various CTAB concentrations. Both lines are displaced to higher concentrations when the temperature is increased (Fig. 51). However, the lines stay parallel to each other. This means that at different temperatures the "degree of occupation" in order to reach one of the limits is exactly the same. But an increase of temperature weakens the binding of the ions to the micelles. Thus in order to reach a certain degree of occupation at higher temperature the equilibrium concentration of KCNS must be increased.

We are now able to compare—at constant temperature—the binding of various anions. Fig. 52 shows the concentrations of KJ and KCNS needed to reach the coacervation limit. For KJ too, we find that the concentrations needed are linear functions of the CTAB concentration. The slope of the line is the same as that of KCNS. This is of great importance as it indicates that the degree of occupation of the micelle surface at which coacervation starts is equal for J-ions and CNS-ions. The difference between KJ and KCNS must be ascribed solely to the difference in equilibrium of binding. To reach the same degree of occupation one needs a higher KJ concentration than KCNS (compare the points of intersection with the ordinate, which lie higher for KJ than for KCNS). To put it in other words: the affinity of CNS⁻ for CTA⁺ is larger than that of J⁻ for CTA⁺, thus:

$$CNS > J.$$

When using KNO₃ (and CTAB) it is not possible to reach the coacervation limit (the system becomes elastic). With NaNO₃, which disolves better

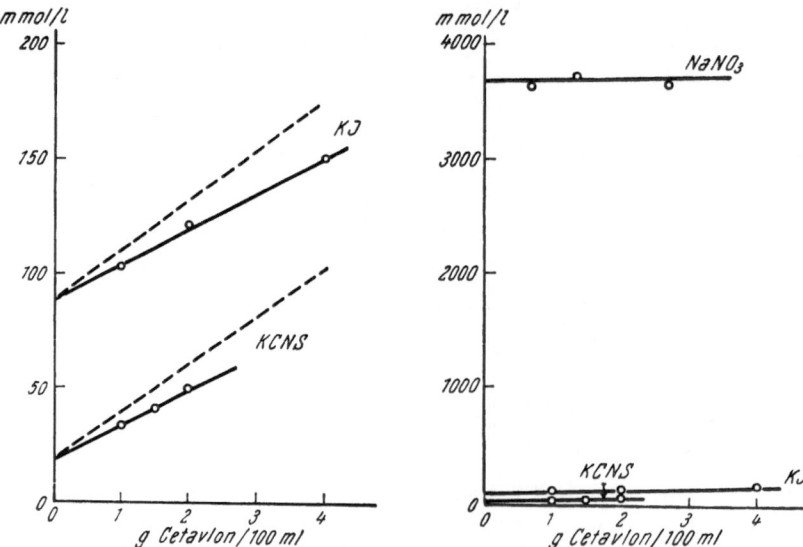

Fig. 52. Concentrations of KCNS and KJ needed to reach comparable coacervate volumes (dotted lines: degree of occupation 100%).

Fig. 53. Concentrations of NaNO₃, KJ and KCNS needed to reach comparable coacervate volumes (CTAB).

in water, it is possible to coacervate a CTAB solution, but the concentration needed is very high. We may conclude that the affinity of the NO_3^--ion for CTA⁺ is much less than that of CNS⁻ or J⁻. When the coacervation limit is determined as a function of the CTAB concentration we find a horizontal line (Fig. 53). Plotted in this scale the influences of KCNS and KJ too, seem to have a horizontal course. We expect that the slope of the NaNO₃ line is equal to that of the KJ and KCNS lines, but this slope is very small in relation to the very high NaNO₃ concentration needed. The expected differences are smaller than the experimental error.

The influence of KBr is less than that of NaNO₃; a saturated solution does not coacervate the system, it will only show elasticity. With NaCl or KCl even the elastic properties are very weak at the highest possible concentration. The sequence representing the affinity of the anions for CTA⁺ appears to be:

$$CNS > J > NO_3 > Br > Cl.$$

Compare chapter 3 (Fig. 17), where we found the same sequence for the affinity of these anions for the positive groups of proteins.

With the anionic detergents the amounts of salt needed are so high as to exclude the possibility of measuring the degree of occupation (in this

Salt	Elastic limit (moles/l.)	Coacervation limit (moles/l.)
LiCl	(precipitate)	(precipitate)
NaCl	0.28	(precipitate)
KCl	0.40	1.7
RbCl	0.55	2.2
CsCl	0.70	2.5

case we are dealing with the bound cations) in the same way as was done in the CTAB/KCNS combination. Here we have the 'same unfavourable situation as in the case with CTAB/NaNO$_3$. The concentrations needed in order to reach the elastic limit or the coacervation limit are for all practical purposes equal to the equilibrium concentrations of the free ions in the medium. From this it follows that we are able to determine experimentally the sequence of affinity of the cations from the respective coacervation limits.

For oleate BUNGENBERG DE JONG, LOEVEN, and WEIJZEN (1950) found the following concentrations of various chlorides needed to reach the elastic limit and the coacervation limit. Thus we find the affinity sequence (see table above):

$$Na > K > Rb > Cs.$$

In these case of LiCl and NaCl complications arise which make it impossible to measure the elastic limit or the coacervation limit directly. Yet it is possible to determine the place of these salts in the affinity

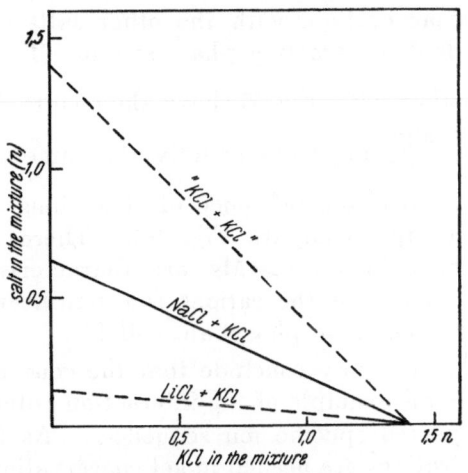

Fig. 54. Concentrations of mixtures of LiCl + KCl and NaCl + KCl needed to reach a certain coacervate volume (oleate).

sequence indirectly by combining these salts with KCl. We recall to mind one of the rules of ion-antagonism (p. 23); when two salts do not differ very much in activity one will find an additive behaviour in mixtures of these salts. Thus BOOIJ [15] determined the relative activities by measuring the coacervate limits in mixtures of LiCl + KCl and NaCl + KCl (Fig. 54). The sequence Li > Na > K clearly comes to the fore. In the range of the elastic systems BUNGENBERG DE JONG, LOEVEN, and WEIJZEN

---

[15] Not yet published.

(1950) found the sequence $Na > K$ for the point of minimum damping (compare next chapter).

Summarising all direct and indirect measurements we find the following sequence:

$$Li > Na > K > Rb > Cs.$$

Teunissen (1936) had already shown that the same sequence comes to the fore when the reversal of charge of oleate is measured (as very high concentrations are needed, he was only able to measure the influence of LiCl, NaCl and KCl).

The results mentioned support strongly the idea that for the appearance of elastic systems or coacervates in solutions of detergents the binding of the oppositely charged ions to the micelle is of essential importance.

As we know that the specific sequences vary for colloids containing different negative groups [16] it is very interesting to study the sequence of the cations when we are dealing with a detergent with a sulfate group. Bungenberg de Jong and Mallee (1952) investigated the coacervation of "Teepol" (a secondary alkylsulfate) with several cations. As the same complications arise which we had already encountered with oleate systems, the investigation had to be done along the indirect way. Only $MgCl_2$ gives coacervation, with the other salts crystallisation or the separation of a double refracting phase sets in. Thus $MgCl_2$ has been combined with the other salts. Fig. 55 shows the results. The sequence of affinity $\left(K > \dfrac{Ca}{2} > Na > \dfrac{Mg}{2} > Li\right)$ shows exactly the same properties which have been found as regards the influence of these ions on macromolecules containing sulfate groups (compare Fig. 16). There too the sequences $K > Na > Li$ and $Ba > Sr > Ca > Mg$ are characteristic. Moreover the influence of the valency of the cations is insignificant as compared with the case of the carboxyl or phosphate colloids.

We may conclude that the coacervation of carboxyl and sulfate soaps is an example of the interaction colloid anion + micro cation with the expected specific ion sequences. As in most cases very high salt concentrations are needed to get coacervation in detergent solutions one is tempted to describe this phenomenon as a simple "salting out" [17]. Yet it must be stressed that one should speak of complex coacervation (of the type colloid anion + micro cation), where the micelle is the polyvalent colloid anion. The elastic-viscous systems are considered to be complex sols of

---

[16] See p. 18.

[17] Under simple "salting out" we will understand those cases in which the addition of a sufficient amount of salt causes the medium to grow too polar to hold a dissolved substance in solution. As a typical example we mention the salting out of alcohol from a water/alcohol mixture by K_2CO_3. Two layers appear: 1. a liquid rich in alcohol and poor in salt and water and 2. a liquid rich in salt and water and poor in alcohol.

the type colloid anion + micro cation. We remind of the definition of the term complex sols (a liquid in which the presence of complex relations is clearly indicated, but in which the intensity of these relations is not strong enough to lead to the separation of a new phase).

The general idea that the elastic-viscous systems and the coacervates of detergents are to be seen as complex systems neglects the fact that the micelle (the "polyvalent colloid anion") is in itself a variable system. Now we may take one step further and say that *the micelle of the association colloids is a complex system.* This complex system is of molecular dimensions and contains as essential components the—by virtue of their carbon chains—associated detergent ions and the oppositely charged ions. Thus this complex system belongs in the class micro anion + micro cation, be it that this is a very special case of this category.

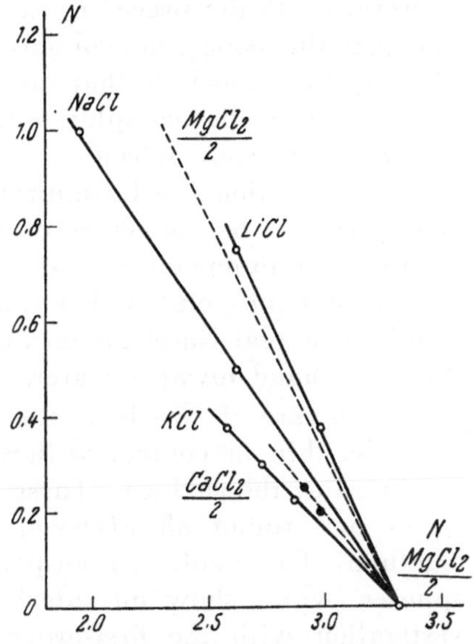

In the absence of salt and in diluted solution practically no complex relations exist in the micelle. The ever proceeding binding of oppositely charged ions to the ionised groups of the detergent at increase of the salt concentration has the following effect:

a) change in shape: spherical micelle → (cylindrical micelle) → sandwich micelle, b) increase of the number of detergent ions participating in the micelle (growth in the plane of the sandwich micelle), c) increased parallelism of the carbon chains (more compact packing), d) arising of spots at the surface of the sandwich micelle which lead to interaction with other sandwich micelles.

Fig. 55. Determination of the cation sequence for the coacervation of T-pol by means of combining the investigated salts with MgCl$_2$.

The four groups all arise from the increasing complex relations between the constituents of the micelle and we may call these relations: *intramicellar complex relations.*

It will be clear that the complex character of the macro systems (elastic-viscous systems and coacervates) are due to *intermicellar complex relations* (see group d in the survey given above). Their intensity will be much lower than that of the intramicellar complex relations. We are dealing here only with the rest of the forces; those not already used up for the intramicellar interactions. Though they are of a different order of magnitude, the intermicellar complex relations are of the same nature as the intramicellar complex relations. Thus the ion sequences will be the same for the elastic limit, the coacervation limit and for the intramicellar complex relations.

d) Phosphatides regarded as association colloids

In chapter 4 we have already seen that phosphatides may participate in various types of complex systems (uni-, di- and tricomplexes). The systems treated there were macroscopic complexes (complex sols, coacervates or flocculations). In order to explain the influence of pH, of salts, etc. on these systems it was satisfactory to take into consideration the amphoteric nature of the phosphatides. In uni- and tricomplex systems both oppositely charged groups take part in the complex relations, in dicomplex systems the excess negative charge is the most important factor.

Refering to the preceding section it may be said that phosphatides too belong to the association colloids and consequently they may form micelles. We may then conclude that the phenomena treated in chapter 4 are based on the intermicellar complex relations or on the complex relations between micelles and macromolecules.

In this section the intramicellar complex relations of the phosphatide molecules will be the centre of our interest. These intramicellar complex relations are in our opinion of the utmost importance for biology.

Our starting-point will be the big difference in behaviour between the simple detergent Na-oleate on the one hand and the amphoteric lecithin on the other hand towards water.

In the case of Na-oleate the substance dissolves spontaneously. If we bring lecithin in contact with water double refracting myelin tubes are produced at the surface. These tubes come off at a slight agitation of the water and round off afterwards, or they roll up while the coils fuse together. The result is a suspension of microscopical, somewhat angular spheres, which show an interference figure in polarised light. An investigation with the first-order-red plate shows that the little spheres consist of concentric layers in which the carbon chains are placed perpendicular to the surface of the sphere. Thus the structure is comparable to that of the myelin tubes (see Fig. 5), in which the concentric layers are cylindrical. The rounding off must be ascribed to the interfacial tension, to which the smectic phase can respond in consequence of its softness. Thus before the rounding off the smectic phase consists of parallel layers with the phosphatid molecules perpendicular to the surface. These layers are separated by water layers.

The conspicuous difference between Na-oleate (dissolves spontaneously) and lecithin (forms a smectic phase) is clearly a result of the fact that in Na-oleate both ions are free from each other as a consequence of electrolytic dissociation, while in the lecithin molecule both charged groups are forced to stay at approximately the same distance. Na-oleate dissolves in water under the formation of spherical micelles. An important part of the Na^+-ions will be found in a diffuse double layer and intramicellar complex relations will be practically absent (Fig. 56 a). The structure of the lecithin molecule (fixed $+$ and $-$ charges at short distances from each other) is ideal for the formation of intramicellar complex relations (Fig. 56 c). Lecithin molecules may be arranged easily into a sandwich micelle, as then the intramicellar relations appear to full advantage. Nothing hinders the outgrowth of the bimolecular layer to quasi infinite dimensions. As a result of intermicellar complex relations of the same type, but of weaker intensity, the double layers will attract each other and a three dimensional structure comes to the fore. From these considerations it will be clear that lecithin—when in contact with water—gives a smectic phase directly.

The simple detergents (e. g. oleate) will only give a smectic phase if enough salt has been added to the medium (Fig. 56 b). The difference between Figs. 56 a and b is that in the first case the intramicellar complex relations are practically absent, while they are in full maturity in the second case [18].

We draw attention to the fact that the Figs. 56 b and c resemble each other not only as regards the parallelism of the carbon chains (perpen-

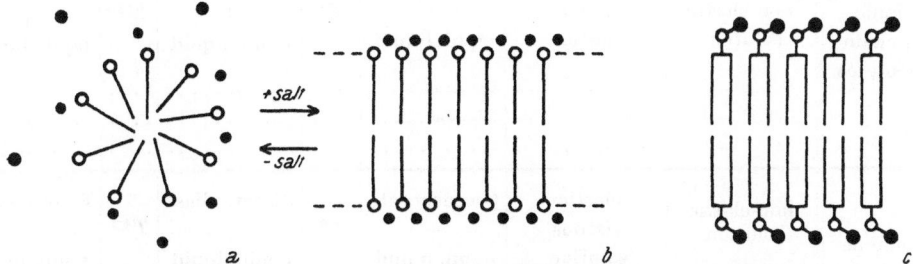

Fig. 56. A Comparison of oleate sandwich micelles (b) and lecithin micelles.

dicular to the double layer), but also as regards the pattern of the + and − charges. The smetic phases of simple detergents and phosphatides accordingly resemble each other in many respects [19]. The difference between Figs. 56 b and c lies in the nature of the complex relations. The simple detergent (Fig. 56 b) shows intramicellar complex relations of the dicomplex type, those of the phosphatides are of the unicomplex type (Fig. 56 c).

This difference is of much practical importance for biology as will be explained in the next section. Here we will rather draw attention to the similarity of the charge pattern in the Figs. 56 b and c. In both cases the ratio of + and − charges is 1 : 1. This can be expressed by saying that the positive and negative charges compensate each other.

In the case of anionic and cationic detergents we arrive at the smectic phase by gradually compensating the charge of the amphipatic ions in the surface of the micelles; whereby we pass sucessively the elastic-viscous systems and the O-coacervates (Compare section c). When start-

---

[18] Compare section b of this chapter and Fig. 41.

[19] A. Similar growth phenomena of the smectic phase (myelin tubes) and comparable optical properties. B. Formation of P-coacervates (with striking properties at the interface coacervate/equilibrium liquid) under the influence of polar non-electrolytes (see p. 130).

ing from the smectic phase (high salt concentration), we will go in the reverse direction if we decrease the salt concentration, viz. smectic phase—O-coacervate—elastic-viscous systems—non-elastic solution. This means a decompensations of the electric charge.

We may now ask whether O-coacervates, elastic-viscous systems and non-elastic solutions may be realised with a lecithin. When we do not permit extreme changes in the pH the answer will be in the negative. As the ratio of charges is 1 : 1, and as these charged groups are bound

Table 4.

	Dry colloid + H$_2$O ↓ non-elastic solution	Elastic-viscous solution	O-coacervate + equil. liquid	T^*	Smect. phase + equil. liquid	T^{**}	P-coacervate + equil. liquid
Anionic or cationic long-chain electrolytes					→ · · · · · · · · · · →	· · · · →	
			→ · →				
		→ · →					
Amphionic long-chain electrolytes	non-elastic solution ← ─ ─	elastic-viscous solution	O-coacervate + equil. liquid	T^*	Smect. phase + equil. liquid ─ ─ ─ ↑ dry colloid + H$_2$O	T^{**} · · · ·	P-coacervate + equil. liquid · · · →

T^* Transition system (smectic phase + O-coacervate + equilibrium liquid).
T^{**} Solution or transition systems of various kinds.

together by covalent bonds, there is no simple means to arrive at a sufficient decompensation of charges. Indeed the above named colloid systems of phosphatides were hitherto never met with.

The only conceivable way to arrive at a decompensation of lecithin is to bind extra charges to the micellar surface. It has recently been found that there are two ways to arrive at this end, a) the binding of an ion with very pronounced affinity to one of the charged groups of the phosphatide, and b) the incorporation of anionic or cationic detergents into the phosphatide micelle. Only the first way will be discussed here, the second will be discussed later (see chapter on elastic-viscous systems).

It has been found (Bungenberg de Jong, van den Berg, and Weijzen 1955) that the salicylate ion has a very distinct affinity for the quarternary ammonium group of cetyltrimethylammoniumbromide. As the main constituent of egg lecithin (phosphatidylcholine) contains this quarternary ammonium group, it seemed worth while to investigate the influence of Na-salicylate on suspensions of the smectic phase of egg lecithin. It has been found (Bungenberg de Jong, de Bakker, and Andriesse 1955) that at increase of the Na-salicylate concentration the phosphatide suspension is transformed successively in a coacervated system, an elastic-viscous system and finally in a non-elastic solution. The coacervate showed all the properties of the O-coacervates of simple detergents, likewise the elastic-viscous system clearly belonged to the same class as those characteristic of simple detergents. The results are theoretically important as the close correspondence of the colloid chemistry of anionic and cationic association colloids (upper row in Table 4) on the one hand, and of amphionic association colloids (lower row of the survey) on the other hand is clearly demonstrated.

The upper row is the same as has already been given at the end of section 2. We remind that horizontal solid arrows mean increase of the concentration of a suitable salt, or—now expressed more generally—compensation of the charge of the micelle.

The broken arrow in the lower row shows the effect of increasing concentration of Na-salicylate, which means here decompensation of the charge ratio 1 : 1 present in the smectic state of the phosphatide. The two vertical arrows give the simple entry (colloid + water) to the two rows of colloid systems.

The dotted arrows give the action of a suitable higher alcohol. It is seen that in order to get P-coacervates the co-operation of a salt is necessary in the upper row, but is not necessary in the lower row (BUNGENBERG DE JONG, DE BAKKER, and ANDRIESSE 1955).

We may close with the remark that no principle objections can be made when for certain aims (e. g. actions of organic non-electrolytes) we study the so much easier obtainable colloid systems of detergents (e. g. oleate coacervates in chapter 7) than those of phosphatides. For other aims (e. g. which are related to specific ion influences on the charge pattern) this, of course, does not apply.

e) Suitability of phosphatides for biological purposes

The presence of phosphatides in the enclosing membranes at the surface of, or within protoplasm has exceptional advantages. Indeed, such membranes might be built up from simple amphipatic molecules (e. g. fatty acid)—see Fig. 56 b—but then serious difficulties might be expected. In the first place fatty acid anions are not possible at pH = 7 (already at pH = 9 hydrolysis sets in). Secondly the realisation of a continuous (dicomplex) membrane would require a certain—often very high—salt concentration. In the supposed case of a membrane of a quaternary ammonium soap (e. g. CTAB) the presence of a very exceptional anion (e. g. CNS^-) would be necessary. These requirements for the realisation of bimolecular layers based on intramicellar dicomplex relations are not met with, neither in the environment, nor in protoplasma itself.

The biological creation of phosphatides solves the difficulties mentioned at one blow. As the charged groups in the lecithin molecule are fixed by covalent bonds at a short distance from each other, the conditions for intramicellar (uni-) complex relations are fulfilled (Fig. 56 c). Hence the living cell does not need—neither in the medium nor in its protoplasm—a highly concentrated or exceptionally composed salt mixture.

In the biological range of pH-values pure lecithin is characterised by the presence of two practically completely ionised groups (pK$_a$ of esterified phosphoric acid $\sim$ 1.5 and pK$_a$ of the quaternary ammonium group $\sim$ 12). Thus in a double layer of pure lecithin the unicomplex relations are maximal and the situation is—in the biological range—not influenced by variations in pH. Moreover, salts like NaCl or KCl will have practically no influence on the unicomplex relations (in consequence of their weak affinity for the charged groups of lecithin). If they have any influence, it will be a decrease of the complex relations (presumably mainly the intermicellar relations).

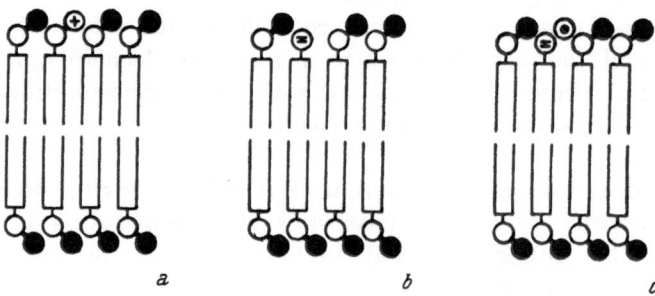

Fig. 57.   Electrical decompensation in phosphatide micelles *a* A cephalin molecule in between lecithin. *b* Decompensation by phosphatidic acid. *c* Ca-ions may compensate phosphatidic acids.

The living cell, however, must have the possibility of—according to the circumstances—increasing or decreasing the complex relations. It will be clear from the foregoing that this is possible only when there is no complete compensation of the positive and negative charges.

The intermediary metabolism has the disposal of ways in which a big or small excess of negative charges may be brought into the double layer reversibly:

a) Transmethylation processes permit the change of the NH$_2$-group of colamin into the quaternary ammonium group of lecithin. In principle lecithin and cephalin may be converted into each other. The NH$_2$-group of cephalin, however, has a lower pK$_a$ than the quaternary ammonium group of lecithin. Consequently the pH range in which cephalin is present as an ampho-ion is shorter than that of lecithin. At pH = 7 already the dissociation of the NH$_2$-group begins to decrease. Thus, if a biochemical conversion of lecithin into cephalin is possible *in situ* (that is to say without the molecules leaving the double layer), then a certain decompensation would be the result and the structure of the double layer would become looser (see Fig. 57 a).

b) By hydrolysis of the bond choline/phosphoric acid a decompensation would result too. The rest of the lecithin molecule which remains after hydrolysis of this bond is phosphatidic acid. Now the decompensation has much more effect than in case a. In case a the change will be felt especially at high pH values. At pH $= 7$ the change is not very big, as there practically all NH_2-groups will be ionised. The splitting off of choline, however, changes the situation drastically. One positive charge per molecule is lost, while moreover one negative charge is gained (at least at sufficiently high pH). We must expect that the compactness of the carbon chains in the double layer decreases much more than in case a (Fig. 57 b).

One may suppose that intermediary metabolism, when need arises, may enzymatically esterify choline to the phosphate group or perform the reverse reaction. Then an internal regulation of the compactness of the double layer is possible. In this view the protoplasmic membranes become subjected to an internal regulation [20].

The hydrolysis of the choline/phosphoric acid bond in the lecithin molecule means that the double layer changes into an ion-exchanger. The choline-ions split off may be exchanged for other cations. Of course, the affinity of these cations for the phosphate group determines the consequences as regards the compactness of the micelle. The addition of Ca^{++}-ions e. g. may tighten a double layer of lecithin containing phosphatidic acid effectively (Fig. 57 c). Here again we may point to the problem of the antagonism of the ions (see p. 23) and the rôle of cations in permeability (see chapter 9).

The considerations treated up to here all refer to the simplest type of phosphatides (lecithins, cephalines, sphingomyelins, acetal phosphatides), viz. those in which one acid group is found beside one basic group. There exist still other types. An interesting type is—especially in relation to the preceding reflections—phosphatidyl serine. Here we find two acid groups (phosphate and carboxyl) beside one basic group (amino). Thus in this molecule there exists no longer a compensation between the $+$ and $-$ charges. As here the ratio between the $+$ and $-$ charges is such that we would expect—for a simple detergent and the same ratio—an elastic-viscous system, the supposition might be expressed that the elastic-viscous properties of protoplasm might be ascribed to double layers of this or a comparable phosphatide. We will expect also that a certain admixture of phosphatidyl serine to a double layer of lecithin will contribute to the decompensation in a comparable way as phosphatidic acid.

We will close this section with the remark that we have treated only one aspect of the double layers of phosphatides. Several other aspects (the presence of organic non-electrolytes—cholesterol—etc.) will be treated in the chapter on "biological membranes."

References

BERG, H. J. VAN DEN, 1953: Stromingsdubbelbreking van elastisch visceuze oleaat-systemen. Thesis, Leiden.

BOOIJ, H. L., 1949: Chapter XIV in KRUYT's Colloid Science II, Amsterdam.

[20] Compare the experiments on *Allium cepa*, p. 147.

Bungenberg de Jong, H. G., and A. de Bakker, 1955: Contributions to the knowledge of P-coacervates IV and V. Proc. Kon. Ned. Akad. Wetensch. Amst. B 58,
   321—354.
— — and D. Andriesse, 1955: Contributions to the colloid chemistry of phos-
   phantides I and II. Proc. Kon. Ned. Akad. Wetensch. Amst. B 58, 238—265.
— H. J. van den Berg, and W. W. H. Weijzen, 1955: Viscous-elastic systems of
   cetyltrimethylammonium bromide and sodium salicylate. Proc. Kon. Ned.
   Akad. Wetensch. Amst. B 58, 133—159.
— and L. J. de Heer, 1949: Soap coacervates with special properties, hitherto
   only known in coacervates of phosphatides. Proc. Kon. Ned. Akad. Wetensch.
   Amst. 52, 783—800.
— W. A. Loeven, and W. W. H. Weijzen, 1950: Elastic viscous oleate systems
   containing KCl. XIII. Proc. Kon. Ned. Akad. Wetensch. Amst. 53, 1122—1135.
— and C. Mallee, 1952: Contributions to the problem of the association between
   proteins and lipids. IV. Proc. Kon. Ned. Akad. Wetensch. Amst. 55, 360—372.
— and A. Recourt, 1953: Binding of the oppositely charged salt ions to the soap
   ions in the formation of elastic systems and in coacervates. Proc. Kon. Ned.
   Akad. Wetensch. Amst. I 56, 303—314; II 56 315—323, and III 56, 442—450.
— und R. F. Westerkamp, 1932: Die Autokomplexkoazervate des Lecithins und
   ihre Bedeutung für das Permeabilitätsproblem. Biochem. Z. 248, 335—374.
— and W. W. H. Weijzen, 1951: Soap coacervates with special properties, hitherto
   only known in coacervates of phosphatides. III. Proc. Kon. Ned. Akad.
   Wetensch. Amst. 54, 81—90.
Debye, P., 1949: Light scattering in soap solutions. Ann. N. Y. Acad. Sci. 51,
   575—592.
— and E. W. Anacker, 1951: Micelle shape from dissymmetry measurements.
   J. Phys. Coll. Chem. 55, 644—655.
Hess, K., H. Kiessig, und W. Philippoff, 1941: Über weitere Ergebnisse röntgeno-
   graphischer und viskosimetrischer Untersuchungen an Seifenlösungen. Fette
   und Seifen 48, 377.
Teunissen, P. H., 1936: Lyophiele niet amphotere bio-kolloiden beschouwd als
   electrolyten. Thesis, Leiden.

6. Elastic-Viscous Systems of Association Colloids

a) Introduction

All three kinds of long chain electrolytes—a) long chain anions (e. g.
oleate), b) long chain cations (e. g. cetyltrimethylammoniumbromide) and
c) long chain amphions (e. g. phosphatides)—may under suitable conditions form elastic-viscous systems. These systems are clear and homogenous solutions, which show a number of characteristic properties (anomalous viscous behaviour, elasticity, streaming birefringence, thread pulling
properties). They are very interesting systems from the biological point
of view, as the above mentioned properties may also be found in protoplasm. Most authors ascribe these properties in protoplasm to protein
systems, but it might be supposed that they are due to our third category
of association colloids (phosphatides), which are present in protoplasm in
abundance.

Unfortunately, experimenting with phosphatides is very difficult for
many reasons. Thus after it had been found at our laboratory once by
chance some ten years ago, that phosphatides may give elastic-viscous
systems, it was thought worth while to study the analogous viscous
systems of simpler association colloids.

Especially elastic-viscous systems made from oleate + KCl (or other
K-salts) have been studied in our laboratory. At the moment the influence

of various factors (salt concentration, detergent concentration, temperature, addition of organic substances) on the four characteristic properties (anomalous viscosity, elasticity, streaming birefringence, thread-pulling properties) is fairly well known.

It is not our intention to give in the present chapter a condensed survey of the work done on oleate systems. We will select a few topics—in particular concerning the viscosity and the elastic behaviour—which are not characteristic for oleate, but will be found with elastic-viscous systems of long chain electrolytes of the other two categories as well. At the end of this chapter we will say a few words on the elastic-viscous systems of phosphatides, which are being studied at present.

b) Viscosity

We have already seen that an oleate solution changes into an elastic-viscous system at certain salt concentrations. These systems combine the properties characteristic for a liquid (streaming at the slightest shearing stress) and of a solid body (elasticity). We take e. g. an oleate solution (1.2%) in $1n$ KCl (at high pH). At 15^0 C. this system is highly viscous and little air bubbles enclosed in the system will oscillate if one gives the vessel a rotational impulse. At first sight the system gives the impression of a clear gel. On closer investigation it appears that the term gel is out of place. A real gel is a system which will be reversibly deformed by a force only when the shearing stress does not pass a certain critical value—the so-called yield value. With the aid of the method of MICHAUD (1923) the question of the presence of a yield value in the elastic-viscous soap systems may be studied. The apparatus (Fig. 58) consists of two wide tubes connected by a capillary. The oleate system is levelled by opening the stop-cock in the wider tube connecting the vessels. After closing the stop-cock a glass rod can be moved up or down in one of the vessels. Then slight differences of level result. The velocity of displacement in the capillary is measured under the microscope (very small droplets of paraffin oil have been added to the system). If a yield value should be present, one would obtain a diagram like Fig. 59 a or b, when plotting the velocity of displacement against the shearing stress. In a certain zone (two times the yield value) no streaming occurs (in this zone we find elastic deformation). In the absence of a yield value Fig. 59 c or d will apply. The experiments on oleate systems correspond to case c in Fig. 59, compare Fig. 60. This proves that the systems begin to flow at the slightest shearing stress; *it is not possible to demonstrate a yield value* (BUNGENBERG DE JONG and VAN DEN BERG, 1949). Equally important is the fact that a straight line is obtained. This fact proves that in this realm (small shearing stresses) the elastic-viscous system behaves like a *normal (Newtonean) liquid*.

At much higher shearing stresses the flow becomes abnormal. Fig. 61 shows the resulting flow rate curve, obtained by the method of PHILIPPOFF

(1936).   As both P and V have been plotted logarithmically, a straight line with a slope of 45⁰ means that the liquid behaves like a Newtonean liquid (velocity of flow proportional to the shearing stress).   The curves for H_2O, 60% sucrose and a calibration oil show that these three liquids behave like Newtonean liquids. The curve for the elastic-viscous system takes a

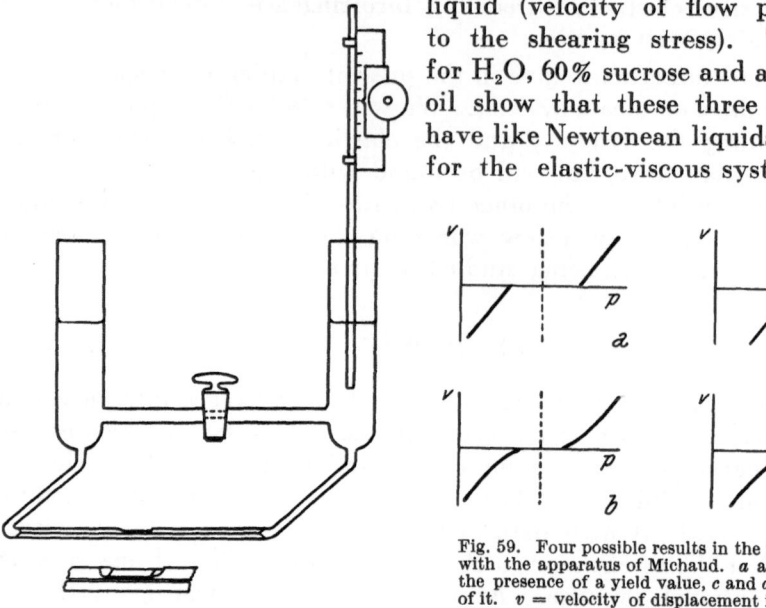

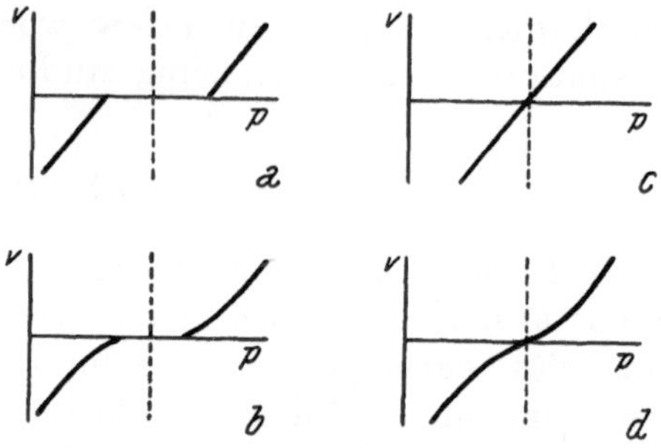

Fig. 59.   Four possible results in the investigation with the apparatus of Michaud. a and b indicate the presence of a yield value, c and d the absence of it. v = velocity of displacement in the axis of the capillary, p = position of the rod. The exact levelling occurs at a value of p indicated by the dotted vertical lines.

Fig. 58.   Apparatus used to investigate the presence or absence of a yield value.

totally different course.   The branch at the extreme left has a slope of 45⁰; in this range of—low—shearing stresses the system has the character

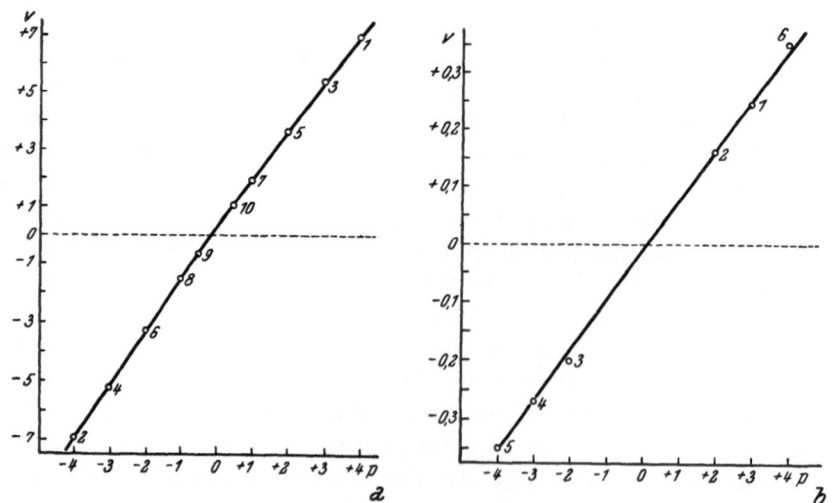

Fig. 60.   Absence of a yield value and occurrence of Newtonean flow of a 0.45% (left) and a 1.14% (right) elastic-viscous oleate system in the range of small shearing stresses. Maximal shearing stress at the wall of the capillary at $p = +4$ or -4 is approximately but 0.07 dynes/cm².

of a Newtonean liquid (compare Fig. 60).   At a certain increase of the shearing stress, however, a strong decrease in viscosity comes to the fore.

At the extreme left of the curve the viscosity is about 10.000 times that of water; at the extreme right it is only 4 times as high. Compare also Fig. 62, where log P/V (that is to say log viscosity) is plotted against log P. The left horizontal part of the curve represents once more the behaviour of the elastic-viscous system as a normal Newtonean liquid: the viscosity η is independent of the shearing stress. When one determines at one and the same system the shear modulus G

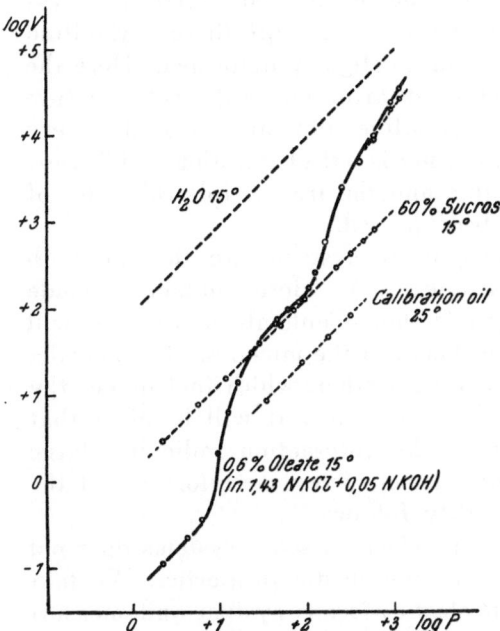

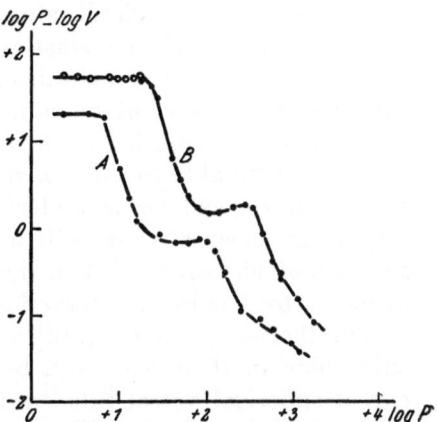

Fig. 61. Flow rate curve for a 0.6% elastic-viscous oleate system. $V (= 4Q/\pi R^2) =$ mean velocity of flow expressed in cc./sec. $P (= R \cdot p/24) =$ shearing stress at the wall of the capillary expressed in dyne/cm².

Fig. 62. Log of the viscosity (log P — log V) of a 0.6% (curve A) and an 1.2% (curve B) elastic-viscous oleate system as a function of log of the shearing stress.

and the relaxation time λ, it appears that the product $G \cdot \lambda$ is of the same order of magnitude as the viscosity η of the left hand level in Fig. 62 (Bungenberg de Jong, Van den Berg, and De Heer 1949). This means that the viscosity in the range of low shear stresses—where the system behaves like a Newtonean liquid—is determined chiefly by the elastic properties, in conformity with the relation given by Maxwell: $\eta = G \cdot \lambda$.

c) Elastic-viscous systems are in thermodynamic equilibrium

One of the most characteristic properties is that the changes in viscosity under the influence of the shearing stress are *completely reversible*. At each shearing stress a certain viscosity is found, which viscosity is established immediately.

The real gels show a completely different behaviour, as may be demonstrated e. g. by a gel obtained by cooling down a dilute gelatin solution (Bungenberg de Jong, Van den Berg, and Verhagen 1952). If we now pass the yield value the gel will begin to flow, but the change is irreversible. The whole history of the system (time of the experiment, the magnitude of the shearing stress) determines the resulting flow. Thus we arrive at the conclusion that the internal situation of an elastic-viscous oleate system and a dilute gelatin gel is fundamentally different. In both cases we must assume a three dimensional mesh-work of mutually cohering elements (because of the elasticity), but these elements as well as the way in which they make contact with each other are completely different.

In the gelatin gel the elements are macromolecules, connected at "crystalline spots." A low shearing stress does not disrupt these crystalline spots; the gel structure remains intact, but is slightly deformed. Here the gel behaves like a solid body. At higher shearing stress the gel structure is destroyed and the chance that the crystalline spots are restored is very small. A restoration of the original situation is out of question. This view elucidates the existence of a yield value and the irreversible character of the system after the yield value has been passed.

The elements of the elastic-viscous oleate systems are the sandwich micelles (perhaps the cylindrical micelles too). Here contact is made between the surfaces of neighbouring micelles. Coulomb interactions will be the main contribution to the contact between the micelles. The micelles will always be able to shift in respect to each other, which fact makes the absence of a yield value understandable. Moreover it will be clear that very large shear stresses will annihilate this interaction radically (large decrease of viscosity). When the system is set at rest, a restoration of the structure by Coulomb interactions quickly follows.

The thermodynamic equilibrium of the elastic-viscous systems does not only show in their viscosity, but also in the elastic properties. We take e. g. a spherical vessel filled with an elastic-viscous system and measure the period of oscillation (T) and the damping Λ (logarithmic decrement). Now this vessel is rotated vigorously in a conical track. Then the system rotates in one direction in respect to the glass wall. The vessel is suddenly placed on a cork ring. The first seconds the system continues to rotate with decreasing velocity, then it stops and starts its damped oscillation. In spite of the vigorous motion and flow, a measurement of T and Λ gives exactly the same values one has obtained previously.

Finally it may be added that the changes in the properties of the birefringence of flow of elastic-viscous systems are also completely reversible.

d) Characteristic quantities describing the elastic properties

The period (T) and the damping (Λ) depend in general on the shape of the vessel and the type of oscillation. Both should be chosen in such a way that it is possible to compute the characteristic quantities in a simple way. The most simple shape of the vessel is the sphere and the most simple type of oscillation is that in which motion takes place in concentric spherical layers or shells.

Burgers (1948) has given the theory of this oscillation. He arrives at the formula:

$$T = \frac{2\pi R}{4.49}\sqrt{\frac{\varrho}{G}}$$

(T = period in sec., R = radius of the spherical vessel in cm., ϱ = density and G = shear modulus in dyne/cm²). This formula shows that T is proportional to the radius R. This relation is always borne out experimentally

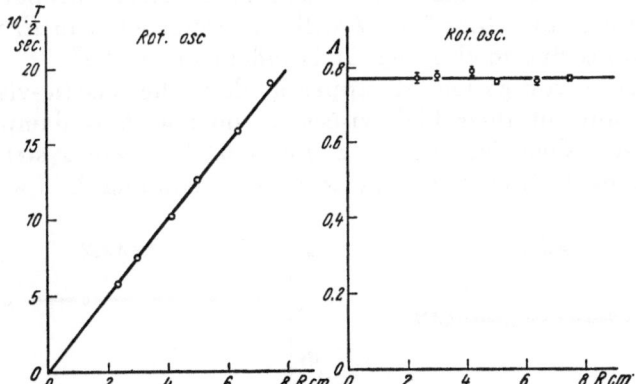

Fig. 63. T and Λ of an 0.6% elastic-viscous oleate system as a function of R.

(see Fig. 63 and 64). Burgers has distinguished three possible causes for the damping of the oscillations of a spherical mass of an elastic fluid: a) purely viscous damping, b) damping through relaxation of elastic

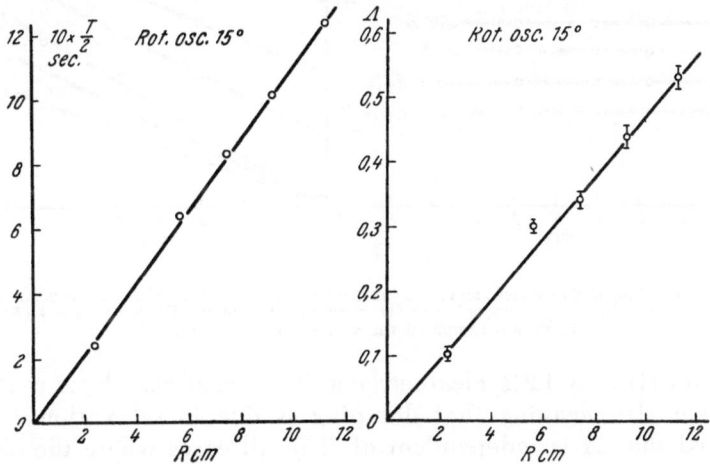

Fig. 64. T and Λ of an 1.2% elastic-viscous oleate system as a function of R.

tensions, characterised by a single, constant relaxation time λ and c) damping through slipping of the fluid along the wall of the vessel. These three types of damping may be described by the following formulae:

a) $\Lambda = \dfrac{4.49\,\pi\,\eta}{R\,\sqrt{G\varrho}}$ (damping due to viscous forces; η = viscosity),

b) $\Lambda = \dfrac{\pi\,R}{4.49\,\lambda}\sqrt{\dfrac{G}{\varrho}}$ (damping due to relaxation),

c) $\Lambda = \dfrac{2\,\pi}{4.49}\dfrac{\sqrt{G\varrho}}{\varkappa}$ (damping due to slipping; $\varkappa$ = friction coefficient).

The formulae show that it will be possible to determine the cause of damping by measuring Δ in spherical vessels of varying radius. Indeed, in the case of viscous damping Δ will be inversely proportional to R, while in the case of relaxation Δ will be proportional to R, while in the case of damping due to slipping, Δ is independent of R.

On closer investigation it appears that the elastic-viscous oleate systems—in spite of their high viscosity—do not show damping through viscous forces. One does find e. g. for a 0.6% oleate system that Δ is independent of R (Fig. 63), that is to say damping is due to slipping

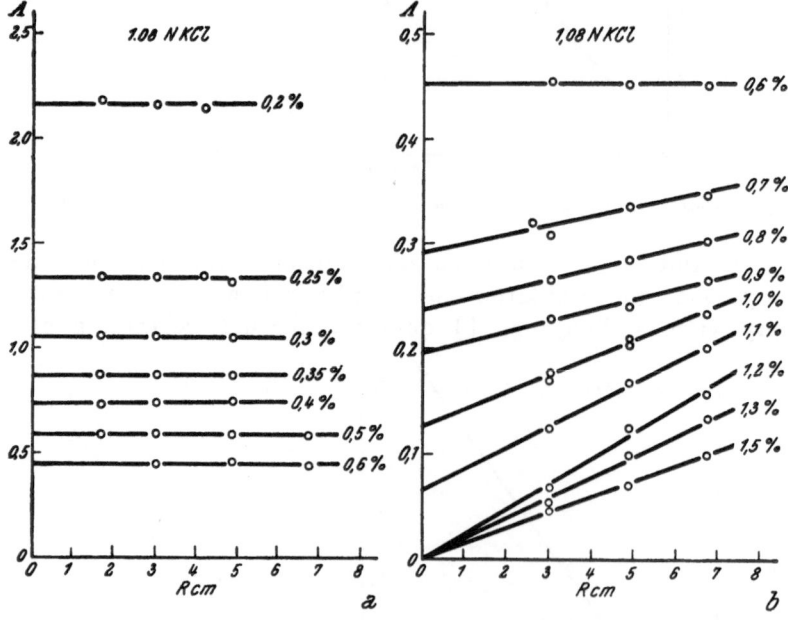

Fig. 65. Λ as a function of R at increasing oleate concentration (% = g./100 ml.). KCl concentration = 1.08 n (that is the KCl concentration corresponding to the minimum damping). From the $\Lambda = f$ (R) curves in this figure are computed the $\varkappa$ and λ values in Fig. 68.

(publication III). A 1.2% oleate system is characterised by $\Delta \sim R$ (Fig. 64) (publication II); meaning that damping is due to relaxation. Moreover it appeared that Δ is independent of R in all cases where the oleate concentration is smaller than 0.6% (publication XIX). When this concentration is higher than 1.2%, Δ is proportional to R. In the intermediate zone (0.7–1.1%) Δ is a linear function of R (Fig. 65). In this intermediate range the two causes of damping (relaxation and slipping) are active simultaneously. The two quantities λ and $\varkappa$ may be computed after a preceding division of Δ in to two parts: 1. $\Delta_{\mathrm{ind.}}$ (Δ independent of R, given by the point of intersection of the Δ line and the ordinate) and 2. $\Delta_{\mathrm{prop.}}$ (given by $\Delta - \Delta_{\mathrm{ind.}}$). With the aid of the formulae given above $\varkappa$ may be computed from $\Delta_{\mathrm{ind.}}$ and λ from $\Delta_{\mathrm{prop.}}$

A single remark might be added to this section. Notwithstanding a great many experimental trials it has not been possible to find slipping of the system along the glass wall in the case of Δ independent of R (Fig. 63) (publication XIV). Thus we must assume that slipping occurs within the elastic-viscous system itself.

e) G, ϰ and λ as a function of the oleate concentration

(Publication XIV)

The elastic properties of an elastic-viscous oleate system depend on many factors (concentration of oleate, of salt and temperature). If one wants to measure the influence of one factor it is necessary to keep the other factors constant. In Fig. 66 it is shown how the shear modulus G and the logarithmic decrement Λ depend on the salt concentration (spherical vessel of constant radius; $R = 6.78$ cm.). It is seen that the G curve has the shape of an S, while the Λ curve is characterised by the presence of a minimum. This minimum lies—when varying the oleate

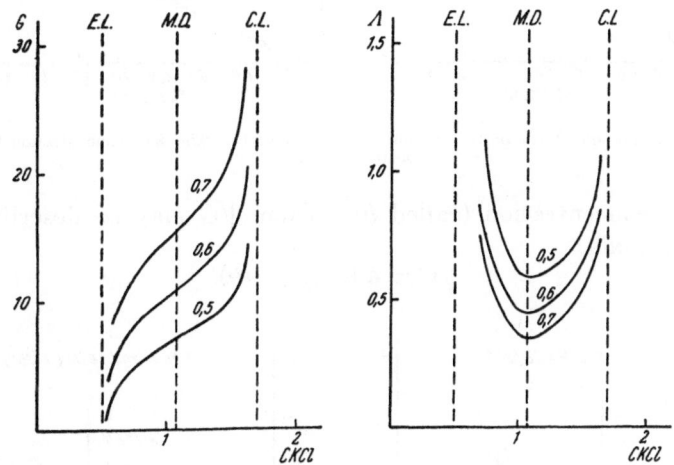

Fig. 66. Shear modulus (G) and logarithmic decrement (Λ) as a function of the KCl concentration at a few oleate concentrations (g./100 cc.). Dotted vertical lines: elastic limit (left), salt concentration of minimum damping (middle) and coacervation limit (right).

concentrations—at practically the same salt concentration. So we may speak of a KCl-concentration of minimum damping (dotted line in Fig. 66; M.D.). The inflexion points of the G curves lie also at the salt concentrations of minimum damping. Because of the fact that G as well as Λ may be determined most accurately at this point, we will choose the KCl concentration at minimum damping for the investigation of the influence of the oleate concentration. It is not possible, however, to compute $\varkappa$ and λ from Λ measurements in one single spherical vessel with known radius. Thus it is necessary to measure the damping in a series of spherical vessels with various values of R. The results of this investigation (oleate concentration 0.2%–1.3%) are plotted in Fig. 67 and 68.

G increases enormously with the oleate concentration. Compare also Fig. 66, where the increase of G at oleate concentrations 0.5–0.7% is pictured. A very simple relation comes to the fore when we plot $\sqrt{G}$

against the oleate concentration (Fig. 67). All points determined experimentally lie on a straight line, which intersects the abscissa at a certain

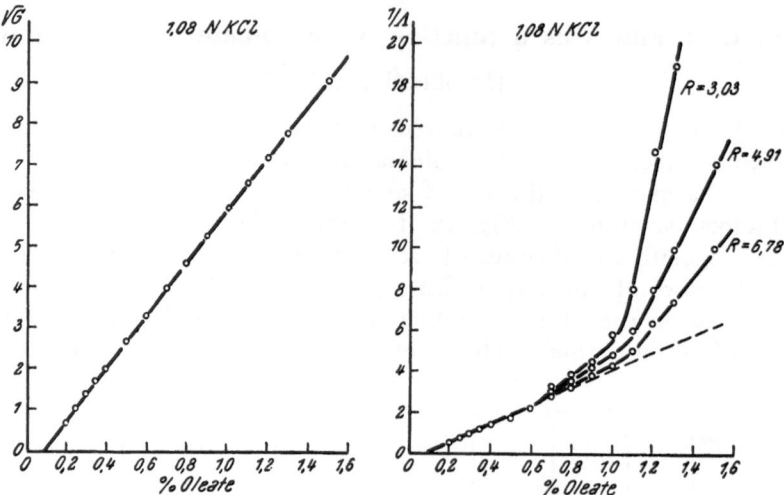

Fig. 67. $\sqrt{G}$ as a function of the oleate concentration (g./100 ml.). The KCl concentration has been kept constant = 1.08 n.

finite oleate concentration (called b). Then $\sqrt{G}$ may be described by the
following formula:

$$\sqrt{G} = a\,(C_{\text{oleate}} - b).$$

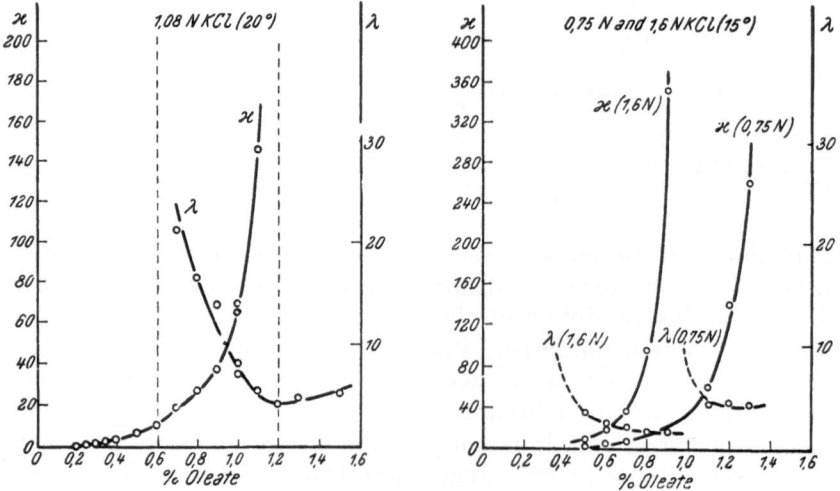

Fig. 68. λ and $\varkappa$ as a function of the oleate concentration (g./100 ml.). The KCl concentration has been kept constant = 1.08 n. λ and $\varkappa$ have been calculated from the results in Fig. 65 as has been discussed at the end of section d). The two dotted vertical lines divide the oleate concentrations in three parts. In the left part the damping is only due to slipping, in the middle part to slipping and relaxation and in the right part only to relaxation.

In this formula a is a factor of proportionality (in Fig. 67 the slope of the
straight line).

We have already seen from experiments on streaming birefringence that the sandwich micelle is the characteristic associate in the elastic-viscous systems (compare chapter 5). It is self-evident to assume that interaction of sandwich micelles causes a secondary association and that this secondary association gives rise to the elastic behaviour. In this case we must expect that G depends on a higher power than 1 of the concentration of the sandwich micelles. Thus we are led to the supposition that $(C_{\text{oleate}} - b)$ represents the concentration of the sandwich micelles, as in that case it follows: shear modulus $\sim$ (concentration of sandwich micelles) [2].

The quantity b then represents the oleate concentration which is not present in the sandwich micelles, but which

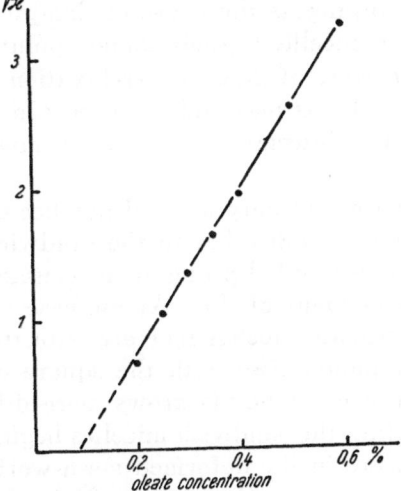

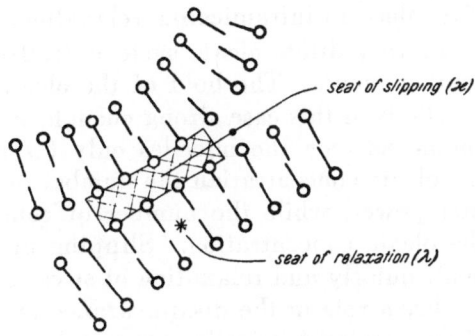

Fig. 69. $\sqrt{\varkappa}$ as a function of the oleate concentration in the region where slipping is the only cause of damping.

Fig. 70. Scheme to explain the two types of damping which may be met with in elastic-viscous systems.

is in equilibrium with these micelles. This oleate might be supposed to be free ions or micelles of another shape (HARTLEY-micelles or cylindrical micelles) [21].

As regards the damping it is seen (Fig. 68) that at first damping due to slipping occurs (characterised by $\varkappa$). At high oleate concentration the cause of damping is exclusively relaxation (λ). In between 0.60% and 1.2% both factors play a role.

Here too we ask ourselves whether the change in the cause of damping at increase of the oleate concentration (slipping→relaxation) is compatible with the supposed structure of the elastic-viscous systems. First of all we will inspect the dependence of $\varkappa$ on the oleate concentration in the range where slipping is the sole cause of damping. In this range (±0.1—0.6% oleate) $\varkappa$ increases strongly with the oleate concentration (the scale of the ordinate in Fig. 68 is too small to observe this clearly). If $\varkappa$ is plotted against the oleate concentration at a larger scale a rapidly rising curve comes to the fore, which curve resembles that of the relation $G = f$ (oleate concentration) very much. If $\sqrt{\varkappa}$ is plotted against the oleate concentration

[21] See p. 42 and Fig. 41.

we obtain once more a straight line (Fig. 69). Here the relation might be formulated as follows:

$$\sqrt{\varkappa} = a_1 \left(C_{\text{oleate}} - b_1 \right).$$

This is the same type of formula as that of the shear modulus (compare Fig. 67). Moreover $b_1 = b$ and we may conclude that $\varkappa$ depends on the concentration of the sandwich micelles in the same manner as G. This fact leads to the supposition that slipping—characterised by $\varkappa$—is localised at the intermicellar contact spots of the sandwich micelles forming the three dimensional meshwork (Fig. 70). Slipping as the cause of damping must be regarded as a tangential shift of the micelles at their contact points.

Then it may be assumed that the other cause of damping—relaxation—must be situated in the micelle itself (Fig. 70). It may indeed be reasoned that at increasing oleate concentration the intermicellar slipping must give place to intramicellar relaxation.

In very dilute oleate systems ($\leq 0.6\%$) we find only a small number of contact points. The hold of the oleate ions to each other in the sandwich micelle is in this case strong enough, so that we find slipping at the contact points between the micelles only (Λ independent of R). At increase of the oleate concentration the number of sandwich micelles increase with the first power, while the number of contact points rises with the square of the oleate concentration. Slipping at the contact points grows more difficult quickly and relaxation of stresses within the sandwich micelles begins to play a role in the disappearance of stresses in the deformed mesh-work of the sandwich micelles (Λ is a linear function of R). At a sufficiently high oleate concentration ($\geq 1.2\%$) there are so many intermicellar contact points that relaxation of stresses in the sandwich micelle has become the sole cause of damping ($\Lambda \sim R$).

f) A comparison of the elastic behaviour of elastic-viscous systems of various soaps

An interesting support of the interpretations laid down in the preceding section has been found in the results of a comparative investigation on elastic-viscous systems of laurate, myristate, palmitate, stearate and oleate (publication XX). This study has been complicated by the fact that a comparison at one and the same temperature is not possible. That is why the dependence on temperature had to be taken into account. Moreover, the salt concentrations (potassiumcarbonate has been used) of minimum damping vary from soap to soap. In the usual way G has been measured as a function of the soap concentration at several temperatures. It appeared that for the other soaps too the formula $\sqrt{G} = a \left(C_{\text{soap}} - b \right)$ holds. It is very interesting to note that the value of a showed to be practically equal for all soaps studied. Compare Fig. 71 (right), where the mean value of a determined at three temperatures has been plotted against the mean of these temperatures. The five points, each representing the value of a for one soap, correspond to a straight line sloping upward weakly to the right. As we know from other investigations that a rises weakly at increase of temperature when working with one soap only, it seems highly probable that the five soaps—if it would be possible to study them at the same temperature— would show the same value of a in the formula $\sqrt{G} = a\,(C_{soap} - b)$.

Obviously, it is of no importance, as regards the shear modulus at the salt concentration of minimum damping, whether the sandwich micelle is built up of soap ions with short or long carbon chains (laurate and stearate) or contains a double bond (oleate). This tallies very well with the

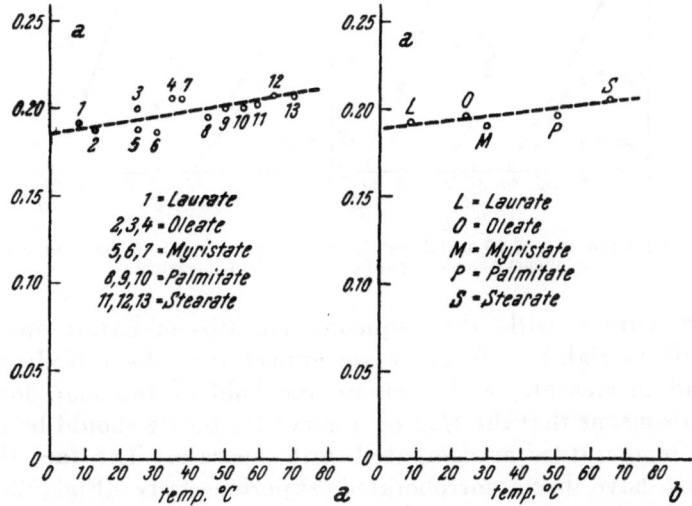

Fig. 71. Values of a in the formula $\sqrt{G} = a\,(C_{soap} - b)$ for a number of soaps, each determined at the individual salt concentration of minimum damping, as a function of the temperature. Left: the individual measurements determined mostly at three temperatures. Right: by plotting the mean of the three values of a against the mean temperature it is shown that a is probably equal for the five soaps (apart from the general slight increase of a with the temperature).

picture developed. Indeed, all sandwich micelles have the same type of surface (ionised carboxyl groups) and G is supposed to be determined by the number of contact points between the surfaces of the micelles.

A comparison of the soaps as regards the damping gives a totally different picture. The investigation would have taken too much time if $\varDelta$ had been measured in many spherical vessels with varying radius. We have at our disposal only the values of $\varDelta$ at one constant R. Besides we have measured $n =$ the number of visible turning points, which is an inverse measure of $\varDelta$ (publication XV). In Fig. 72 $1/\varDelta$ and n are plotted against temperature (soap concentration 50 m mol/l.).

The soap concentration is so high that we are dealing with the zone where relaxation (characterised by λ) is the only cause of damping. In the preceding section it has been suggested that in this case the cause of damping is situated within the sandwich micelle. Then we may expect that relaxation will increase at increase of temperature in consequence of the growing motion of the molecules. Thus we must expect that $\varDelta$ increases at rising temperatures (or $1/\Lambda$ and n decrease; compare Fig. 72). Moreover, we must expect that the hold of the soap ions to each other in the micelle will increase as the carbon chain is lengthened. Thus relaxation will decrease—at the same temperature—when the carbon chain increases. In the diagram giving the relation between $1/\Lambda$ or n and temperature we expect

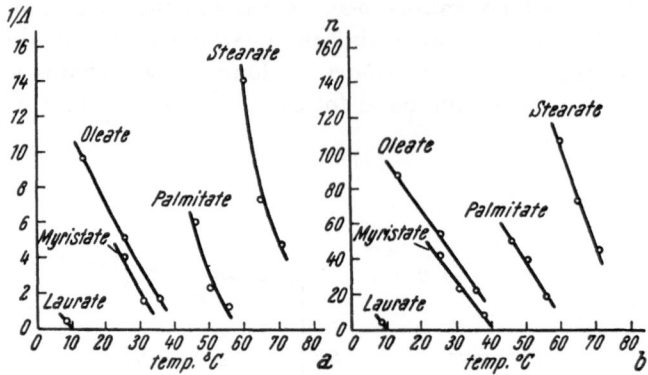

Fig. 72. Values of $1/\Lambda$ and of n of 50 millimoles/l. soap systems at the individual salt concentration of minimum damping as a function of the temperature.

a series of curves with the sequence laurate—myristate—palmitate—stearate (left to right). Moreover we expect that the introduction of a double bond in stearate, will decrease the hold of the soap ions in the micelle. This means that the $1/\Lambda$ or n curve for oleate should be displaced to a lower temperature as compared with stearate. The fact that these expectations have been corroborated experimentally (Fig. 72) gives a strong argument in favour of our hypothesis.

g) Elastic-viscous systems of long chain electrolytes of the second and third categories (phosphatides)

Oleate belongs to the first group of long chain electrolytes (see introductory section of this chapter). Recently we have investigated in some detail the elastic behaviour of an example belonging to the second category (cetyltrimethylammoniumbromide + Na-salicylate, at the salicylate concentration of minimum damping; Bungenberg de Jong, Van den Berg, and Weijzen 1955). In all essential points the results were the same as have already been described for oleate. The same intrinsic change occurs within the micelle (transformation into sandwich micelles) by the binding of oppositely charged ions. Here, of course, we are dealing with the binding of anions to the long chain cations.

In a favourable case (CTAB + KCNS) it was shown that the region of elastic-viscous systems lies at a degree of binding of the anions of 40—70% (see chapter 5). About halfway this region (± half of the ionised groups occupied by anions) the damping is a minimum and consequently here the elastic properties are most easily observed.

Though we cannot give an explanation of this fact yet—minimum damping at half occupation—it gives us a useful hint when trying to realise elastic-viscous systems of phosphatides (our third category of long chain electrolytes). The simplest way seems to be the formation of mixed micelles, containing a long chain electrolyte of the third category (lecithin) and a long chain electrolyte of the first or second group. Compare the survey below, which gives the known sequence of characteristic systems of the first and second category resulting at increased compensation of the charges, and the expected sequence for phosphatides resulting from decreased compensation:

1st and 2nd category — non-elastic solutions → elastic-viscous systems → coacervates → suspensions of smectic phase

→

increased compensation by binding of oppositely charged ions

3rd category — suspensions of the smectic phase → coacervates → elastic-viscous systems → non-elastic solutions

→

decreased compensation by addition of long chain electrolytes of the 1st or 2nd group.

Experiments in this direction have been performed in this laboratory (BUNGENBERG DE JONG an DE BAKKER 1956). They confirm the idea that electrical decompensation of phosphatide micelles (smectic phase) may lead to coacervates, elastic-viscous system and non-elastic solutions in the expected sequence. Returning to our suggestion that phosphatides may contribute to the viscous and elastic behaviour of protoplasm, we think it probable that only those phosphatides play a role, which are sufficiently decompensated electrically [22]. Two possibilities may be mentioned:

a) a phosphatide micelle decompensated by incorporation of phosphatidic acid or b) a phosphatide micelle containing phosphatides with an unequal number of positive and negative charges (e. g. phosphatidylserine).

References

BUNGENBERG DE JONG, H. G., 1951: Elastic viscous oleate systems containing KCl XV. Proc. Kon. Ned. Akad. Wetensch. Amst. 54, 1—12.
— and A. DE BAKKER, 1956: Contributions to the colloid chemistry of phosphatides III and IV. Proc. Kon. Ned. Akad. Wetensch Amst. B 59, 124—161.
— and H. J. VAN DEN BERG, 1948: Elastic viscous oleate systems containing KCl I. Proc. Kon. Ned. Akad. Wetensch. Amst. 51, 1197—1210.
— — 1949: Elastic viscous oleate systems containing KCl II. Proc. Kon. Ned. Akad. Wetensch. Amst. 52, 15—27.
— — 1949: Elastic viscous oleate systems containing KCl III. Proc. Kon. Ned. Akad. Wetensch. Amst. 52, 99—112.

---

[22] Compare p. 61.

Bungenberg de Jong, H. G., and H. J. van den Berg, 1949: Elastic viscous oleate systems containing KCl IV. Proc. Kon. Ned. Akad. Wetensch. 52, 363—376.
— — 1950: Elastic viscous oleate systems containing KCl VII, Proc. Kon. Ned. Akad. Wetensch. Amst. 53, 7—18.
— — 1950: Elastic viscous oleate systems containing KCl VIII. Proc. Kon. Ned. Akad. Wetensch. Amst. 53, 109—121.
— — 1950: Elastic viscous oleate systems containing KCl IX. Proc. Kon. Ned. Akad. Wetensch. Amst. 53, 233—246.
— — and L. J. de Heer, 1949: Elastic viscous oleate systems containing KCl V. Proc. Kon. Ned. Akad. Wetensch. 52, 377—388.
— — W. A. Loeven, and W. W. H. Weijzen, 1951: Elastic viscous oleate systems containing KCl XX. Proc. Kon. Ned. Akad. Wetensch. Amst. 54, 399—420.
— — and H. J. Verhagen, 1951: Elastic viscous oleate systems containing KCl XIX. Proc. Kon. Ned. Akad. Wetensch. Amst. 54, 317—329.
— — — 1952: The elastic behaviour of diluted gelatin gels. Comparison with the elastic behaviour of elastic-viscous oleate systems. Proc. Kon. Ned. Akad. Wetensch. Amst. 55, 1—27.
— — and D. Vreugdenhil, 1949: Elastic viscous oleate systems containing KCl VI. Proc. Kon. Ned. Akad. Wetensch. Amst. 52, 465—478.
— — and W. W. H. Weijzen, 1955: Viscous-elastic systems of cetyltrimethyl-ammonium bromide and sodiumsalicylate. Proc. Kon. Ned. Akad. Wetensch. Amst. B 58, 135—159.
— W. A. Loeven, and H. J. Verhagen, 1950: Elastic viscous oleate systems containing KCl XII. Proc. Kon. Ned. Akad. Wetensch. Amst. 53, 975—988.
— — and W. W. H. Weijzen, 1950: Elastic viscous oleate systems containing KCl X. Proc. Kon. Ned. Akad. Wetensch. Amst. 53, 743—758.
— — — 1950: Elastic viscous oleate systems containing KCl XI. Proc. Kon. Ned, Akad. Wetensch. Amst. 53, 759—774.
— — — 1950: Elastic viscous oleate systems containing KCl XIII. Proc. Kon. Ned. Akad. Wetensch. Amst. 53, 1122—1135.
— — — 1950: Elastic viscous oleate systems containing KCl XVI. Proc. Kon. Ned. Akad. Wetensch. Amst. 54, 240—252.
— — — 1951: Elastic viscous oleate systems containing KCl XVII. Proc. Kon. Ned. Akad. Wetensch. Amst. 54, 291—302.
— — — 1951: Elastic viscous oleate systems containing KCl XVIII. Proc. Kon. Ned. Akad. Wetensch. Amst. 54, 303—316.
— W. W. H. Weijzen, and W. A. Loeven, 1950: Elastic viscous oleate systems containing KCl XIV. Proc. Kon. Ned. Akad. Wetensch. Amst. 53, 1319—1336.
Burgers, J. M., 1948: Damped oscillations of a spherical mass of an elastic fluid. Proc. Kon. Ned. Akad. Wetensch. Amst. 51, 1211—1221.
Michaud, F., 1923: La rigidité des gelées. Ann. de Phys. (Fr.) 19, 63—80.
Philippoff, W., 1936: Zur Messung der Strömung von strukturviskosen Flüssigkeiten. Kolloid-Z. 75, 155—161.

7. The Interaction between Organic Substances and Lipophilic Colloid Systems

a) Introduction

Many non-reactive organic substances have a profound influence on living cells. Long ago it has been found that the action of these substances is correlated with their physico-chemical properties. Their activity increases with the surface activity or with the distribution coefficient lipid/water. In general one might say that the balance between the hydrophobic and hydrophilic groups in the molecule is of the utmost importance for the strength of their biological activity. As the lipids in a living cell consist—for a large part—in phosphatides and cholesterol, it would be interesting to measure the influence of non-reactive organic substances on systems built from phophatides or cholesterol. This is a rather difficult task, which has been attacked from several angles.

In most cases it appeared that phosphatides and cholesterol show practically insurmountable experimental difficulties. Thus, most experiments have been performed with substances which resemble these compounds more or less. One of the oldest examples is the fact that normally the balance between the hydrophilic and the hydrophobic groups of a given substance has been characterised by the distribution coefficient water/olive oil. Olive oil has been chosen as an approach to the natural lipids, but everybody must admit that the distance between the natural lipids and olive oil is rather large. It is, however, quite impossible to measure the distribution coefficient water/cell lipids.

A very useful and interesting method is the study of monomolecular layers. Here too the amount of research on phosphatides and cholesterol is quite small, while other lipids—especially fatty acids—have stimulated a great many experiments. Physico-chemically speaking, the agreement between phosphatides and fatty acids (and other amphipatic molecules) is so great, that one must expect that the experiments on monolayers of fatty acids and related substances will be of high interest to the biologist. The number of experiments on monolayers performed in connection with biological problems is quite large, notwithstanding the fact that the substances forming the films were in general unnatural lipids.

The situation of these molecules in the monomolecular film is far from natural too. It would be interesting to study the phosphatides in their natural medium water. With this idea in mind BUNGENBERG DE JONG started experiments on phosphatide coacervates. The study of these coacervates, which come to the fore only under exceptional circumstances, was far from easy. They contained only a low amount of water and they were very viscous and stuck heavily to the walls of the vessels. That is the reason why he turned his attention to simple models of phophatides: soaps and synthetic anionic and cationic detergents.

b) Methods

It has already been described that the addition of KCl in high concentration to an oleate solution causes the appearance of a coacervate. The influence of organic substances on the coacervate may be demonstrated and measured in several ways.

1. One may isolate a coacervate layer, put a small crystal or a droplet of the substance to be investigated into a drop of coacervate and follow the reaction under the microscope (Fig. 73). This method—the contact method—furnishes a clear classification of organic non-electrolytes (BUNGENBERG DE JONG and SAUBERT 1937).

A. Around the small crystal there is produced a dense field of vacuoles, which expands continuously. In addition we see grains appearing—frequently birefringent—while finally myelin tubes can make their appearance from them. A strong condensing action must be the cause of this phenomenon. The excess of solvent is separated in the form of vacuoles. Addition of KCl would give the same phenomenon; the volume of the coacervate decreases. Examples: benzene, cholesterol and iso-amylurethane.

B. A vacuole field is produced. This is caused by a substance with

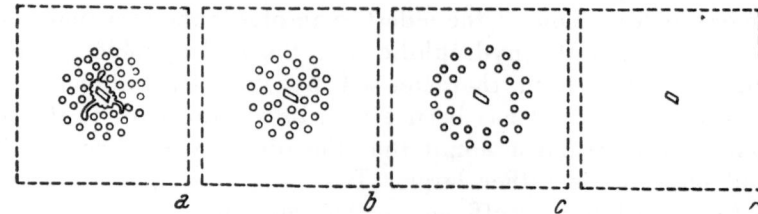

Fig. 73.   Vacuolization of an oleate coacervate under the influence of various substances (Bungenberg de Jong and Saubert 1937).

a weaker condensing action. Examples: phenanthrene, *p*-dichlorobenzene and iso-butylurethane.

C. First we see vacuoles produced around the crystal. This, however, appears to be a ring of vacuoles, which shifts slowly away from the crystal. The crystal finally lies in a clear zone. In this case we must assume that low concentrations of the compound investigated have a condensing action, while on the other hand higher concentrations have an opening action (the uptake of water is favoured). Examples: propylurethane and hydroquinone.

D. Microscopically there is nothing to see. This can have many causes. A strong opening action as well as a weak condensing action would give the same result. Experiments at a higher temperature will make a distinction possible.

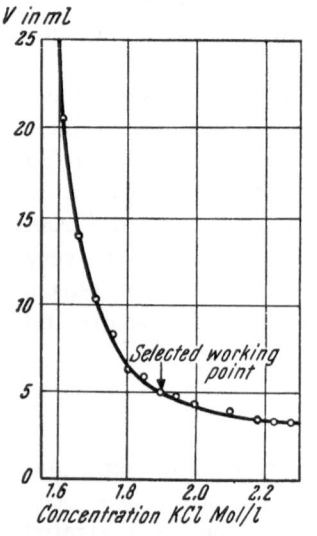

Fig. 74.   Influence of KCl on the volume of the layer of the separated oleate coacervate (Bungenberg de Jong, Booij, and Saubert 1937).

From these experiments it became clear that organic substances may have an opening or a condensing action on the oleate coacervate. As the former action may be counteracted by an increase of the KCl concentration we will give it the neutral name "salt demanding action." The latter is opposed by a decrease of the KCl-concentration, thus we will call it "salt sparing action."

2. Direct measurement of coacervate volumes. A series of graduated stoppered cylinders (volume 50 ml.) is prepared, containing a constant concentration of Na-oleate and increasing concentrations of KCl (Bungenberg de Jong, Booij, and Saubert 1937).

The volumes of the coacervate layers are measured. From a graph (Fig. 74) we select a suitable working point. Then we prepare a new series of cylinders with the constant Na-oleate concentration and varying amounts of a solution of the organic substance to be investigated (the total volume is constant throughout the series). The coacervate volumes sometimes increase under the influence of the added substances (opening or salt demanding action), in other cases the volume decreases considerably (condensing or salt sparing action). Ethylalcohol is an example of the first class of organic substances (Fig. 75 A), while n-butylalcohol belongs to the second class (Fig. 75 B). It will be clear that this method is restricted to substances which are more or less soluble in water.

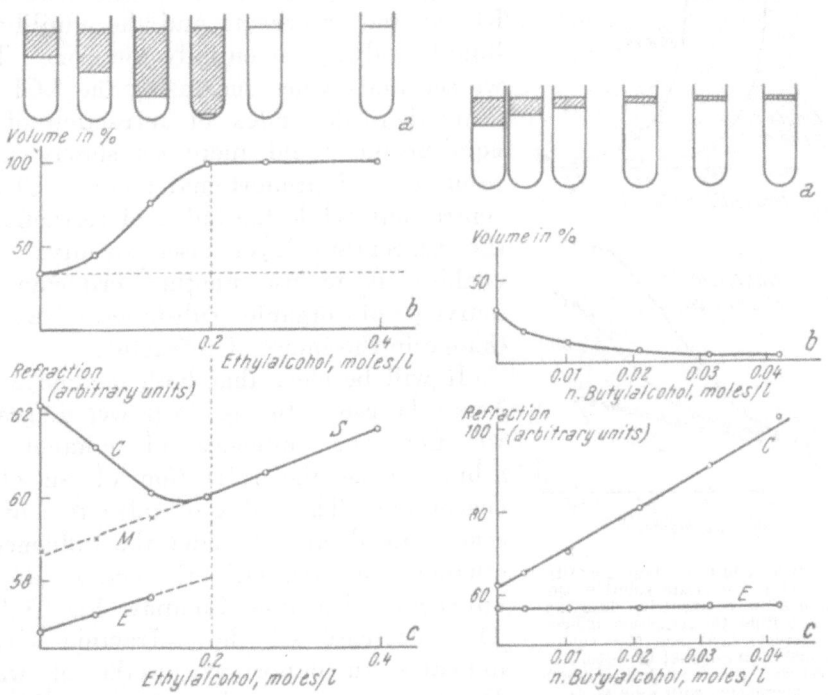

Fig. 75. Left: disappearance of an oleate coacervate (KCl constant) under the influence of ethanol. Right; under the influence of n-butanol the oleate coacervate is more condensed. b Volume of coacervate layers, — c Refraction indices of C = coacervate layers, E = equilibrium liquids and S = non-coacervated solutions (M = "mean refraction").

3. Measurement of the shift of the coacervation curve (BOOIJ and BUNGENBERG DE JONG 1949). Now the organic substance to be investigated is dissolved into the oleate solution. Then we determine the "KCl coacervation curve" of this mixture by adding increasing amounts of KCl and measuring the coacervate volumes (the organic substance now has a constant concentration throughout the series). The curve obtained is compared with the blank oleate curve (Fig. 76). We see that for n-butanol the curve shifts to lower concentrations of KCl (less KCl is needed to give the same degree of coacervation: salt sparing action). The shift of the curve (e. g. at a coacervate volume of 50%) is a measure of the action of the substance. We may experiment with several concentrations of the organic substance and plot the activity of the substance (measured as a shift of the KCl-curve)

against its concentration. This method increases the possibilities of investigation very much as many organic substances insoluble in water dissolve readily in an oleate solution.

4. Refraction of the coacervate and the equilibrium liquid (Booij, Lycklama and Vogelsang 1949). When we add KCl in sufficient amounts to an oleate solution the coacervate layers become ever smaller. With the

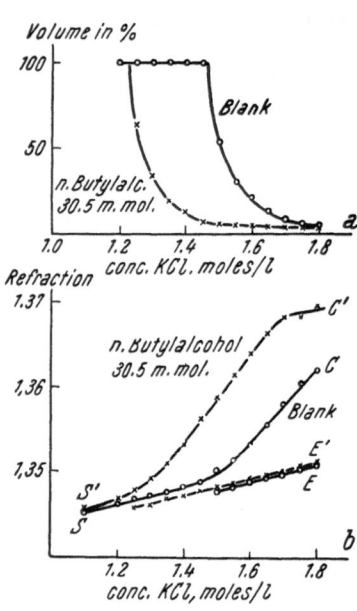

Fig. 76. If we add a constant quantity of n-butanol to an oleate solution, the KCl curve shifts to the left. This can also be seen from the refraction indices (C = coacervate layer and E = equilibrium liquid of the blank; C^1 = coacervate layer and E^1 = equilibrium liquid of the experiments with n-butanol).

aid of a refractometer one may prove that only the oleate is concentrated (condensed) during this process; the concentration of KCl in the coacervate and the equilibrium liquid is always essentially the same. Thus we see that—when increasing the KCl concentration—the index of refraction of the equilibrium liquid increases slowly (as a result of and proportional to the KCl concentration), while the index of refraction of the coacervate layer rises steeply. This enables us to use another criterium for activity of organic substances, viz. the change in the index of refraction.

It will be clear that both method 2 and 3 may be used. In Fig. 75 (lower diagrams) we find the influence of ethanol and n-butanol on the refraction of an oleate coacervate. The index of refraction of the coacervate decreases under the influence of ethanol and eventually the coacervate disappears (broken line). Compare Fig. 75 (left). This decrease of the refraction clearly indicates an important uptake of water. The reverse is shown by n-butanol (Fig. 75, right). The refraction increases strongly.

The shift of the coacervation curve may also be infered from refraction data. This has been shown in Fig. 76. The shift of the KCl curve to the left is clearly indicated.

This method is especially useful when we are dealing with very small coacervate layers. Then it is very difficult to measure the volumes exactly.

c) The activity of aliphatic substances

Using the second method (direct measurement of the coacervate volumes) a large number of organic non-electrolytes have been investigated (Bungenberg de Jong and co-workers 1937—1938). Some general rules could be drawn up as regards their action on the oleate coacervate.

1. In homologous series of compounds with a polar group the concentration at which an effect is attained, decreases with increasing number of carbon atoms.

2. In such a series the action generally reverses at a particular chain length. The lower members have an opening action (e. g. ethanol at high concentrations), the higher ones a condensing action.

3. The influence of compounds with a hydrophilic group seems to consist of two opposite components:

　　a) a condensing action of the hydrocarbon chains,

　　b) an opening action of the polar group.

4. With an equal number of carbon atoms the influence decreases with branching of the chain, with ring closure or on the transition from a saturated six membered ring to an aromatic ring.

5. The introduction of a halogen into an aliphatic chain results in a stronger condensing action.

6. Among the groups which favour an opening action the following might be mentioned: the OH-group, ether-, ketone- and ester-oxygen.

7. In general the action of a compound is determined by the structure of the hydrocarbon chain (number, distribution and state of binding of the carbon atoms) and by the nature, number and position of the polar groups.

The development of the third method (measurement of the shift of the coacervation curve) has given us the opportunity to study more organic substances. Moreover some other substrates have been studied (synthetic detergents).

Fatty acid anions

We thought it worth while to start our new attack on the problem with a series of substances resembling the "substrate" (oleate) very much, viz. the fatty acid anions (Booij and Bungenberg de Jong 1949). The results of a study on the action of fatty acid anions on the oleate coacervate looked rather surprising at first sight (Fig. 77). Short members of this homologous series have practically no influence. Then an opening action comes to the fore, which rises considerably up to the undecylate ion (all experiments are performed at pH $= \pm 12$). After this the activity falls, to rise considerably with the higher fatty acid anions. It should be pointed out that the amount of the fatty acid anions added is much lower than that of the oleate. In an endeavour to explain the curve found we postulated that in an oleate coacervate large flat sandwich micelles are present. In these micelles the carbon chains of the oleate molecules form an ordered structure.

We start from the fact that the addition of a small amount of oleate has practically no influence on the coacervation curve. The reason is clear: a molecule resembling the constituents of a micelle very much does not disturb its orderly structure. We must, on the other hand, expect a disorganising action from a molecule of the same homologous series which is taken up in the micelle, but which has other dimensions. The interesting graph (Fig. 77) may then be explained from the following assumptions:

　　1. In soap coacervates large, flat and orderly built micelles are present.

　　2. The first part of the curve (up to undecylate) is the result of the

unequal distribution of the added fatty acid anions between the medium and the micelles. A short fatty acid anion (e. g. butyrate) is not taken up and consequently has no action on the oleate coacervate (except at very high concentrations). The distribution equilibrium between micelles and medium changes as the terms of the homologous series grow longer. We found—from experiments with various oleate concentrations—that under certain circumstances 19% of the added nonanoate molecules were taken up in the micelles. Under the same circumstances this number was 83% for undecanoate and 100% for stearate. The opening action of undecanoate must be ascribed to the fact that it is taken up in the micelle, where it creates "centres of disorder" as its carbon chain is much shorter than that of the oleate.

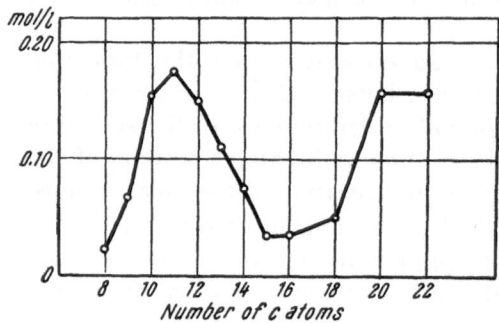

Fig. 77. Activity (shift of the KCl curve, caused by 0.5 m mol. of added substance) of fatty acid anions on the oleate coacervate (Booij and Bungenberg de Jong 1949).

3. The second part of the curve is a consequence of the fact that the carbon chains of the added substances (11–16 carbon atoms) fit better and better in between the parallel oleate molecules. The disturbance of the order decreases. We must remember that the "effective length" of the oleate molecule is certainly less than 18 carbon atoms because oleic acid possesses the cis-configuration. Thus it is not astonishing that we find the minimum of opening action (= greatest resemblance to the oleate molecules) at fatty acid anions with 15–16 carbon atoms.

4. The third part of the curve would be explained by the fact that too long molecules as well as too short ones disarrange the structure of the micelles.

The hypothesis given may be strengthened by several experiments. When experimenting with a stearate coacervate (the temperature must be chosen high—60° C.—; K_2CO_3 is used instead of KCl) we find the minimum of opening action in the homologous series of fatty acid anions at 18 C-atoms. We also performed some experiments on the influence of fatty acid anions on a desoxycholate coacervate. Here the "substrate" does not resemble the added molecules. Thus we do not expect a minimum in the activity of the term of the homologous series. Indeed, we found only a steadily increasing opening action (increasing from the term with 6 carbon atoms up to that with 15 carbon atoms and remaining constant at higher chain lengths).

Undissociated fatty acids

The above experiments had all been performed at high pH. Of course it would be interesting to investigate the influence of undissociated fatty acids on association colloids. We cannot use soap coacervates for this study. When we add an acid to a soap-coacervate (e. g. oleate) we find always the same phenomenon: the volume of the coacervate volume decreases sharply. Inorganic and organic acids all act in the same way. That is why we must assume that oleic acid is formed and that this substance has a strong condensing action. This fact fits in the general theory of the coacervation of association colloids. The flat micelle is the result of an equilibrium between opposite forces. The promoting factor is the attraction between the carbon chains of the oleate molecules. The opposing factor is the repulsion of the negative carboxyl groups (which is decreased by the binding of cations).

The transformation of an oleate ion into an oleic acid molecule must have a drastic condensating effect, as at this place in the micelle the repulsing factor is abolished, while the attracting factor remains in full strength.

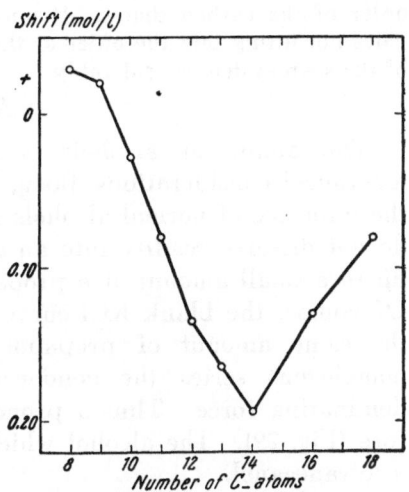

Fig. 78. Activity of fatty acids on a coacervate of dodecylsulfate (shift expressed in mol/l. NH₄Cl; 1 m mol/l. fatty acids added).

In accordance with this view it was found by Booij and Vreugdenhil (1950) that fatty acids have a strong condensing action on a coacervate of dodecylsulfate (at somewhat acid pH). Here too we expect that the influence of the terms of the homologous series will not be equal. A fatty acid which fits exactly in the dodecylsulfate micelle will have the strongest action, as it gives the maximal contribution of the factor which promotes sandwich micelle formation. As the molecule of dodecylsulfate is somewhat longer than the fatty acid with the same number of carbon atoms we are not suprised to see a maximal condensing action at myristic acid (Fig. 78). When using a substrate with a longer carbon chain (a coacervate of cetylsulfate) the maximal condensing activity is found at a correspondingly longer fatty acid (stearic acid). Booij and van Leeuwen (1953), experimenting with the cetylsulfate coacervate, showed that the influence of fatty acids depends very much on the pH. At low values of pH all fatty acids investigated (8—22 carbon atoms) show a condensing activity. An increase of the pH promotes the opening action, especially for the shorter terms of the homologous series. At pH = 8.3 for instance the fatty acids up to palmitic acid show an opening action (maximal at tridecanoate), while the higher acids have a low condensing action.

A condition for the appearance of an outspoken maximum of activity in a homologous series depends—according to our hypothesis—on the ordered structure within the soap micelles. We must expect that we will not find such a maximum when dealing with relatively unordered micelles.

This has been found indeed in experiments on a T-pol coacervate[23] (Booij and van Calcar 1950). As the carbon chains in this substrate vary considerably, the interior of the micelle must be irregular. The activity of the fatty acids on this coacervate slowly increases with increasing carbon chain. No trace of a maximum can be found in this case.

The interior of a micelle consisting of substances with different chain lengths will, of course, contain no vacuum. That means that the longer chains will be folded or spiralized in such a manner that the micelle will show two practically flat surfaces, while the interior is filled up with carbon chains showing a certain degree of disorder. This will influence the properties of the micelles strongly. We have seen e. g. that the damping of elastic soap systems depends a. o. on the order of the carbon chains. It must be expected that the introduction of molecules not fitting into the order of the detergent micelles will increase the damping of the soap system considerably.

Alcohols

The action of alcohols is easily comprehensible from the above-mentioned considerations. Booij, Vogelsang, and Lycklama (1950) measured the influence of normal alcohols on the oleate coacervate. As long alcohols do not dissolve readily into an oleate solution, the substances were taken up in a small amount of n-propanol. The oleate solution was then added. Of course, the blank KCl-curve has been measured after the addition of the same amount of propanol to the blank oleate solution. In this homologous series the condensing action of the carbon chain is the dominating force. Thus a pronounced condensing influence comes to the fore (Fig. 79). The alcohol which seems to fit best in the oleate micelle is n-tetradecanol.

It will cause no wonder that the maximum of condensing action lies at another term of the homologous series when we experiment with another coacervate. Booij and Vreugdenhil (1950) found a maximal condensing action at n-dodecanol when working with a dodecylsulfate coacervate.

It might be possible that these experiments find their counterpart in biology. There too the chain length of non-reactive organic substances is a very important factor. There is, however, one important objection against the idea that the experiments on the influence of alcohols on oleate coacervates are directly related to the biological experiments with the same substances. In the latter one gets the impression that the alcohols exert an opening action on some lipid system in the cell. Perhaps the following experiments may prove to give the clue to this problem (Booij 1952).

---

[23] T-pol is a sec. alkylsulfate with the general formula

$$R_1 - \overset{\displaystyle H}{\underset{\displaystyle OSO_3Na}{C}} - R_2$$

R_1 and R_2 are alkylgroups with a normal chain, R_1 being in the main CH_3 or C_2H_5, while the total number of C-atoms in the molecule varies from 8–18. This soap may be coacervated with $MgCl_2$.

n-butanol is a strongly condensing substance (at least at high pH and acting on the oleate coacervate). When the pH is decreased, however, the condensing activity decreases too and eventually (at pH = 9.3) butanol behaves like a rather strong opening substance. The big difference between an oleate coacervate at pH = 9.3 and pH = 12 ist that the former is much more condensed. Perhaps one might conclude that the more condensed the micelle, the more normal alcohols will tend to behave like opening substances. Then it would follow that most biological lipid

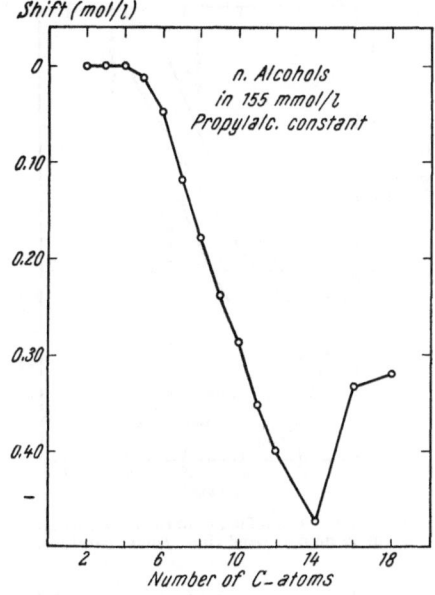

Fig. 79. Condensing action of the normal alcohols (0.5 m mol/l.) on the oleate coacervate.

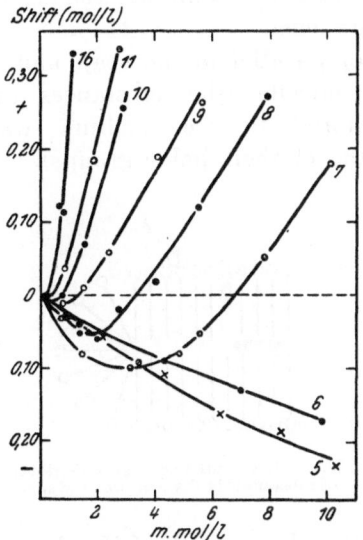

Fig. 80. The action of normal paraffins on an oleate coacervate (+ = opening action, — = condensing action).

systems will be stronger condensed than the oleate micelle (compare e. g. the erythrocyte, where many alcohols exert an opening—hemolysing—action).

Hydrocarbons

An investigation of the influence of aliphatic hydrocarbons on the oleate coacervate (Booij, Lycklama, and Vogelsang 1950) showed that these substances do not fit into the scheme of the action of organic substances as developed in the foregoing pages. One might suppose that the paraffins would show an ever increasing condensing action when ascending the homologous series. The results of the experiments were, however, quite different from these expectations (Fig. 80). The long paraffins show an ever stronger opening action. In an endeavour to explain these unexpected results, we started from two considerations:

1. the paraffins do not contain a hydrophilic group like the substances treated up to now,

2. the experiments with fatty acids etc. show that a disturbance within the micelle (caused e. g. by a too short or too long carbon chain) leads to an opening effect, which is sometimes observed only as a decrease of the condensing action.

It has been suggested that the action of any added substance depends primarily upon its distribution in the soap/water system. Four possible places might be mentioned (Fig. 81):

a) in the medium (practically no influence on coacervation),

b) on the surface of the micelle (here we think especially of ions of the opposite sign, inorganic as well as some organic ions),

c) parallel to the soap molecules in the micelle (the substances are then anchored to the medium—water—by virtue of their polar groups),

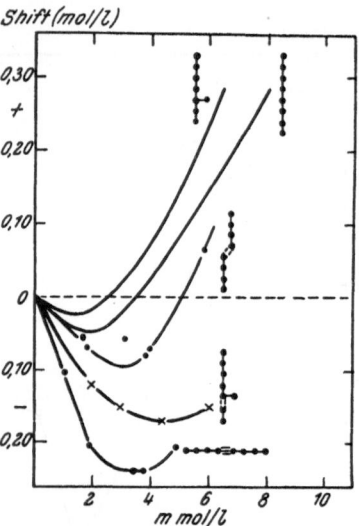

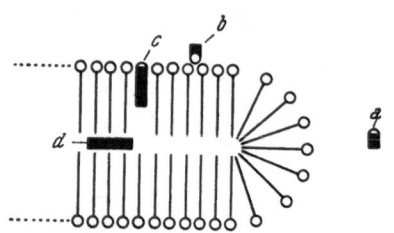

Fig. 81. Possibilities as regards distribution of substances in the soap/water system.

Fig. 82. The influence of the introduction of a double bond into some paraffins.

d) in between the CH_3-planes of the micelle (organic substances without a hydrophilic group or with a weakly hydrophilic group only).

It will be exceptional to find added molecules only at one of these places. n-pentanol e. g. will be found partly at a and partly at c. With n-hexanol the ratio between the amounts found at c and a increases. n-octanol is found practically completely at c (Booij, Vogelsang, and Lycklama 1950).

In the case of the paraffins the experimental results suggest that the distribution between c (condensing action) and d (opening action) depends on the number of carbon atoms and on the concentration of the added paraffin. It is not clear why the shorter paraffins should have a preference for place c.

There have been found several experimental arguments in favour of the suggestion put forward in the preceding paragraphs. Fig. 82 shows the enormous difference in action between 3-methylheptane and 3-methylheptene-2. The latter substance, having a slightly hydrophilic group (the double bond) practically at the end of the molecule tends much more to place c than the former, saturated, substance. Even when the double bond is situated in the middle of the molecule a change of activity is observed, becoming more marked when a triple bond is introduced (compare *n*-octane, trans-octene-4 and octyn-4).

For the biologist it is most interesting that slight changes in molecules of this kind result in marked changes in their action on a relatively simple physico-chemical model. Especially the student of the problem of the relation between molecular structure and physiological action might find some clue for some of his cases, as it is known that the introduction of a double bond into a biologically active molecule may change its activity profoundly.

Halogen-derivatives

The introduction of halogens into paraffins (Booɪᴊ, Lᴄᴋʟᴀᴍᴀ, and Vᴏɢᴇʟ- sᴀɴɢ 1950) means the introduction of a weakly hydrophilic group. Then

the molecule will tend to be anchored in situation *c* (Fig. 81). This means that halogenated paraffins will show a stronger condensing action than the parent paraffins. The experiments gave results which were in complete agreement with these expectations (Fig. 83). From these experiments it is seen that hexane has only a slight condensing action, as compared with bromohexane. Di-bromodecane resembles the higher alkanes, except for the fact that its condensing action is more pronounced.

The introduction of a halogen into the carbon chain of an alcohol, fatty acid, ketone, etc. has a different effect. Then— generally speaking—the action of the parent substance is increased. This is also the result of an influence on a distribution equilibrium, but this time the distribution

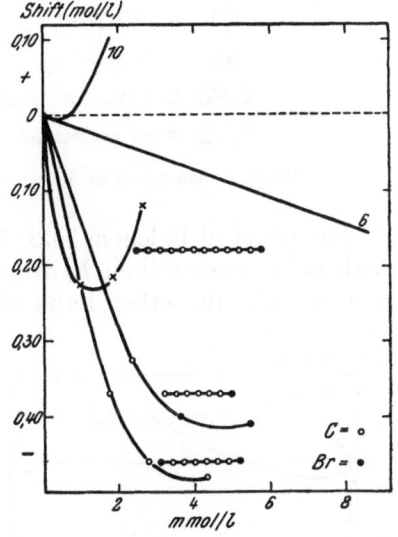

Fig. 83. Influence of the introduction of halogen atoms into alkanes.

between the places *a* and *c* (Fig. 81) undergoes a change. Suppose we have a rather short alcohol (e. g. amylalcohol). In a certain case 50% will be found in the medium and 50% in the soap micelles. When we now introduce a bromine atom at the ω-carbon atom the molecule grows more hydrophobic. The result is that perhaps 75% will be found in the micelles. The condensing activity of amylalcohol is increased.

As the halogens have an influence on the distribution equilibrium water/micelle it is clear that the opening action of other substances (e. g. fatty acid anions) is increased too. Benzoic acid has—at high pH—a slight opening action on the oleate coacervate. Introduction of a chlorine atom at the para-carbon atom of the benzene nucleus increases this opening action considerably. The reason is clear. Benzoate is very soluble in water. Thus only a small amount of the benzoate ions will be found in the soap micelles. They exert—by virtue of their negative groups—an

opening action. Para-chloro benzoate is less soluble in water, thus more
ions will enter the micelles and the opening action increases.

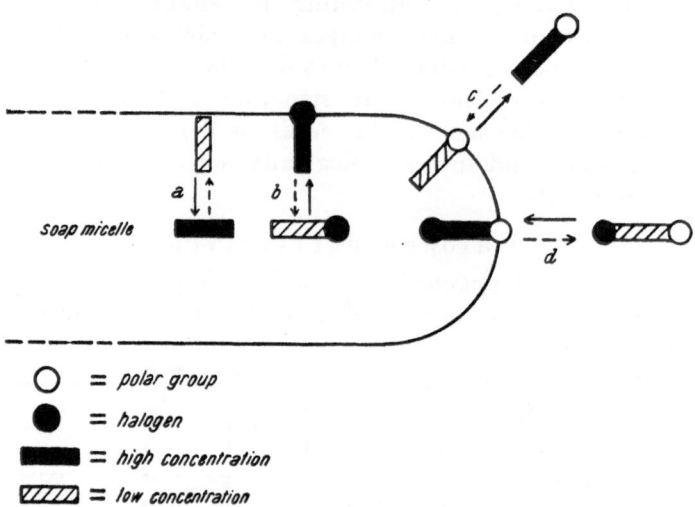

Fig. 84. Influence of the introduction of halogen atoms on the distribution equilibria.

The effect of halogens may be summarised as follows. As we are dealing
with polar groups their hydrophility is larger than that of CH_2- and CH_3-
groups. On the other hand the hydrophility of halogen atoms is much

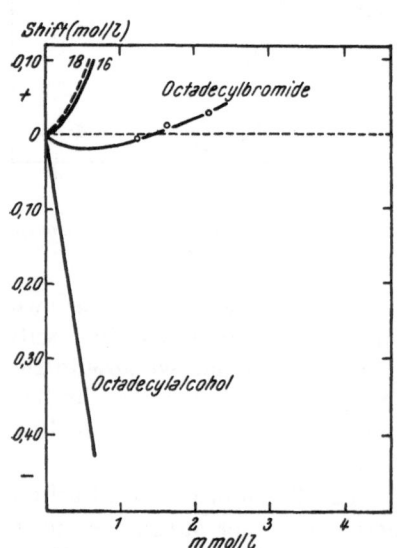

Fig. 85. Influence of the introduction of
polar groups into a long alhane.

less than that of polar groups with ex-
centric dipoles (OH-, NH_2-groups, etc.).
Thus they have no strong tendency to
form hydrogen bonds and consequently
they will be expelled from the water
when possible. Fig. 84 sums up the hypotheses given. A paraffin is practically
not found in the medium, but it is distributed among two places in the micelle
(case a). Introduction of a halogen atom
favours the accumulation parallel to the
soap molecules; the substance is anchored
in this place by its slightly hydrophilic
group (case b). A short alcohol will be
found in the medium and in the micelle
(case c). Introduction of a halogen atom
results in a concentration in the micelle
(case d). From these considerations it
may be deduced that in the case of alcohols and other substances with a strongly
polar group the introduction of halogen atoms does not have an easily
predictable effect, if the parent substances are already taken up completely
in the micelles. Then the distribution equilibrium between medium and

micelle cannot be influenced and properties of molecular structure will be deciding as regards the activity of the substances.

The weak hydrophility of halogen atoms came clearly to the fore in a comparison of octadecylbromide and octadecylalcohol (Fig. 85). The former substance has a slight action only (is distributed among the places c and d of our diagram Fig. 81). The latter is a very strongly condensing substance (its polar group causes a fixation at place c). A long alkyl chain provided with a halogen atom resembles a paraffin much more than a long alcohol does.

Ethers

It will cause no wonder that the action of ethers shows much resemblance to that of paraffins. Fig. 86 (Booij, Kwestroo-van den Bos, and Blekkingh 1954) shows the result of the experiments on the action of ethers on the oleate coacervate. The most conspicuous difference is that the first term which turns from a salt-sparing (condensing) to a salt-demanding (opening) action is somewhat longer for the ether series than for the paraffin series (d-n-amyl-ether versus heptane).

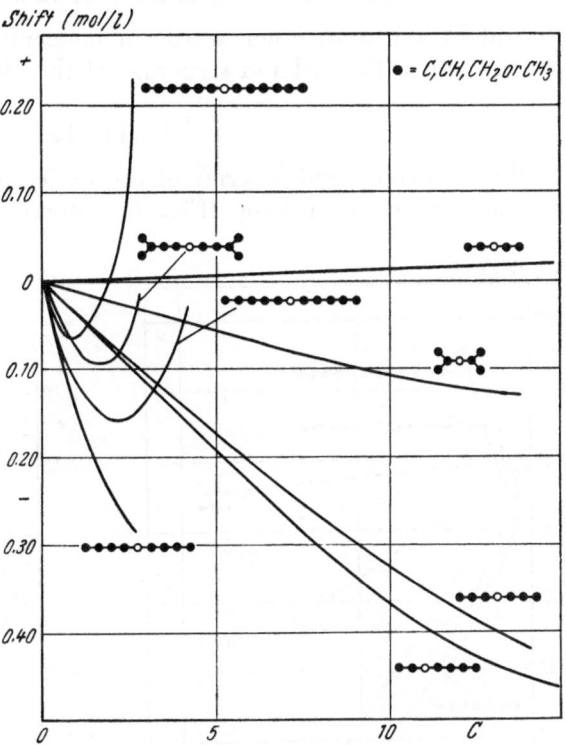

Fig. 86. Activity of aliphatic ethers on an oleate coacervate (abscissa = concentration of the ethers in millimoles/l.).

This tallies with the hypothesis that a hydrophilic group tends to hold the molecule in position c (Fig. 81). Thus in the case of the ethers the apolar part of the molecule must be pretty large to overcome the opposing factor due to the slightly hydrophilic groups.

From Fig. 86 it appears that di-isopropylether has a much lower activity then di-n-propylether. This effect of branched chains has been observed already by Bungenberg de Jong et al. (1938). The distribution between medium and micelle in the case of isomers lies more to the side of the medium for the isomers with branched chains. The same diagram shows that di-isoamylether has a smaller activity than both di-n-amyl-ether and di-n-hexylether. At higher concentrations, however, it takes an intermediate position. This too seems to have a general background. Position c (parallel to the oleate molecules, see Fig. 81) is not favoured by molecules with branched chains. For these molecules the partition equilibrium is shifted somewhat to the position between the CH_3-planes, in the micelles (d in the Fig. 81). Thus we get two different sequences. For the condensing action we find (increasing activity):

di-isoamylether $<$ di-n-amylether.

On the other hand we find for the opening action:

di-n-amylether $<$ di-isoamylether $<$ di-n-hexylether.

We might say it in other words: a molecule with a branched chain does not fit well in the ordered structure of the oleate molecules in the micelle.

Sulfides

We also compared a series of aliphatic sulfides as regards their action on the oleate coacervate (Fig. 87; Booij, Kwestroo-van den Bos, and

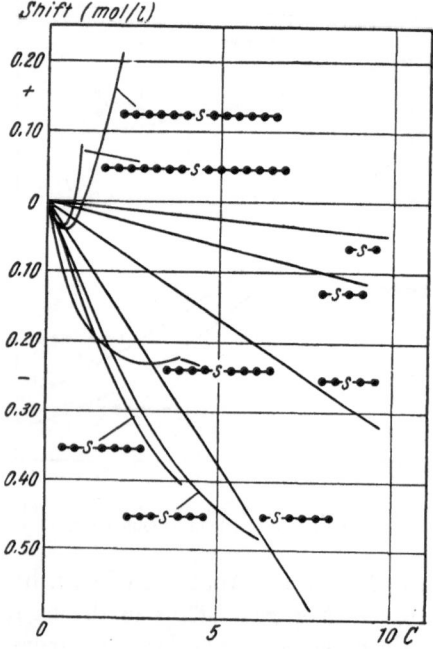

Fig. 87. Action of a series of aliphatic sulfides.

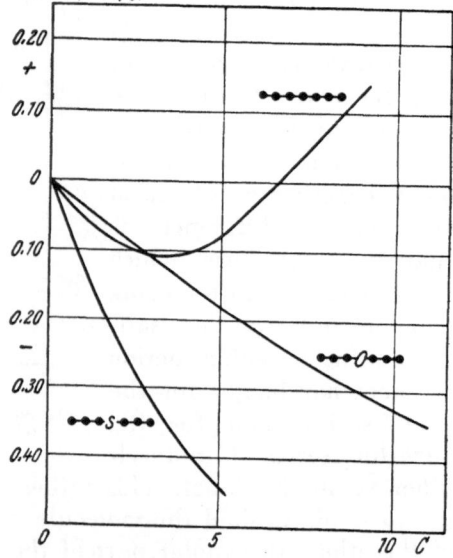

Fig. 88. A comparison of a paraffin, a sulfide and an ether of approximately the same length.

Blekkingh 1954). Generally speaking, the picture resembles that of the ethers and of the paraffins very much. There are, however, typical differences between the ethers and the sulfides, which may be readiliy explained. When we compare e. g. di-n-propylether and di-n-propylsulfide, we see that the latter substance has a stronger salt-sparing (condensing) activity. This must be a result of the difference in the hydrophilic groups, influencing the distribution medium/micelle. The sulfides are more hydrophobic than the ethers.

The same property affects the distribution within the micelle. The —S—

group being a less strong hydrophilic "anchor" of the molecule than the
—O— group, the sulfides will lose their salt-sparing (condensing) activity
sooner than the ethers. This is clearly shown by a comparison of di-*n*-
hexylether and di-*n*-hexylsulfide (Fig. 86 and 87). The difference between
di-*n*-propylether, di-*n*-propylsulfide and "di-*n*-propylmethane" (*n*-heptane)
is shown in Fig. 88. At first sight this is a rather complex picture, but it
becomes clear when we realise that distribution between three positions
(*a*, *c* and *d* in Fig. 81) is involved. Starting from the sulfide in Fig. 88
—which prefers position *c* —we get a tendency to go to *a* of the ether (stronger hydrophilic group) and a preference for position *d* of the paraffin (loss of the hydrophilic group).

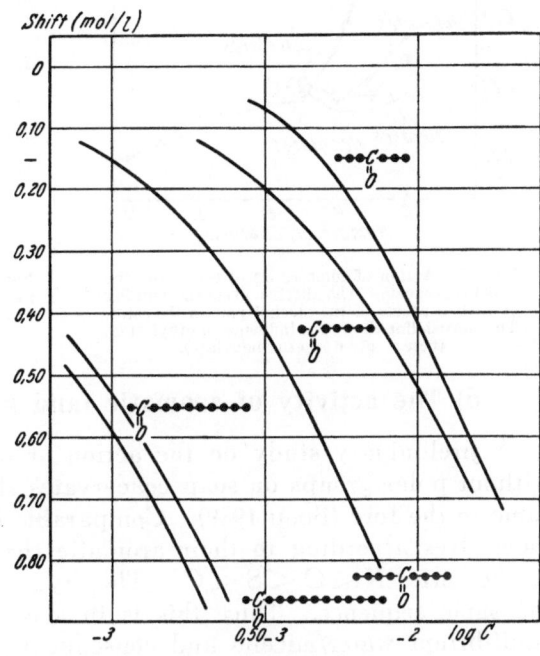

Ketones

The carbonyl group has a hydrophilic character. Thus it causes no wonder that the ketones show an activity which resembles that of the alcohols (BOOIJ, BLEKKINGH, and KWESTROO-VAN DEN BOS 1954). The shorter ketones have an opening action on the oleate coacervate. The longer ones (Fig. 89) show a strongly condensing influence. More-over, the place of the polar

Fig. 89. Influence of some aliphatic ketones on an oleate
coacervate.

group in the molecule appears to have an important influence on its
action. We see that an isomer with the polar group in the centre of the
molecule has less activity than the isomer with an eccentric ketone group
(compare di-*n*-propylketone and methyl-*n*-amylketone). Here too branching of the chains resulted in a decreased activity).

Esters

It is not possible to measure the influence of esters on the oleate coacervate, as these substances are slowly hydrolysed at the pH prevailing in
this coacervate. Thus another substrate (the T-pol coacervate) was chosen
(BOOIJ and VAN CALCAR 1950). The experiments showed an ever increasing
condensing action when ascending the homologous series, provided that the
ester group was situated at the end of the molecules (Fig. 90, acetates and
ethylesters). When, however, the polar group is found in the middle of
the molecules, a shallow minimum comes to the fore (valerates, hexylates).
This minimum is much more pronounced when we compare esters with equal

carbon chains at both sides of the ester group (Fig. 91). It seems that the longer esters show some tendency to concentrate within the micelles, which results in a lower condensing action.

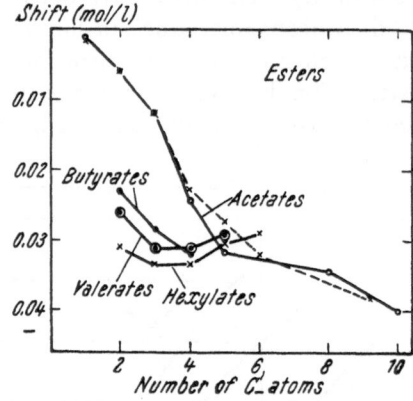

Fig. 90. Action of some series of esters on the T-pol coacervate. The abscissa gives the number of carbon atoms of the alcohol part of the ester. The dotted line shows the influence of ethylesters (here e. g. 6 = ethylhexylate).

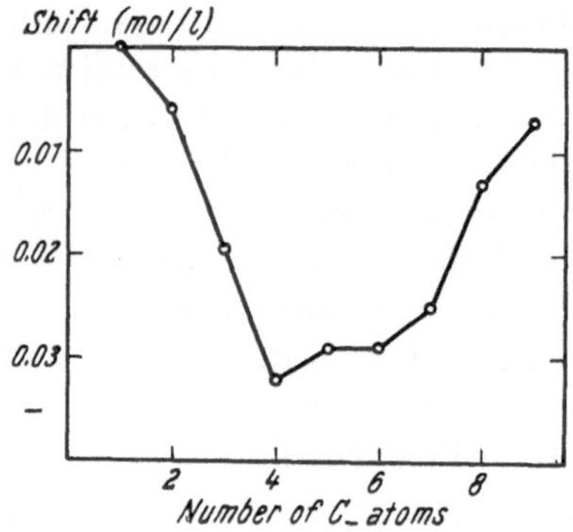

Fig. 91. The action of esters with an equal number of carbon atoms at the acid and the alcohol side of the molecule on a coacervate of T-pol (6 means hexylhexylate, etc.).

d) The activity of aromatic (and heterocyclic) substances

A preliminary study on the action of organic ring systems with and without polar groups on soap coacervates showed that some general rules came to the fore (Booij 1949). Comparable heterocyclic rings will arrange themselves according to their aromatic character, the activity increasing in the series $N < O < S < C$. The hydrophobic character increases in the same sequence. Thus, this is in accordance with the idea that the equilibrium water/micelle and consequently the activity depends on the hydrophilic groups present in the molecule.

The difference between alicyclic and aromatic nuclei is rather complicated, as the action depends on the fact whether or not a polar group is bound to the molecule. Without polar group the aromatic nucleus has a stronger action than the similar alicyclic ring, while with a polar group the alicyclic derivative is the stronger one. In some cases the introduction of a polar group results in a stronger activity than in the parent substance (e. g. naphthol has a stronger condensing action on the cetylsulfate coacervate than naphthalene).

In these cases too, considerations about the distribution of the substances in the soap system help to clarify the problems. First of all it must be stated that the alicyclic rings are more hydrophobic than the comparable aromatic ones. If we have an aromatic ring with a strong polar group (e. g. thymol acting on a cetylsulfate coacervate) we will observe a condensing action. If, however, the benzene ring is replaced by a cyclohexane ring (menthol) the activity is strengthened as the equilibrium water/micelle is shifted to the right. If, on the other hand, no polar group is present the substances will be taken up practically completely in the micelles. Now the distribution equilibrium within the micelle is affected by the change of ring system. An aromatic ring will promote a con-

o-xylene

1, 2-dimethylcyclohexane

centration parallel to the soap molecules, while an alicyclic ring resembles the normal paraffins and part of these molecules will accumulate in the middle of the micelle (weakened condensing activity). Thus we see that o-xylene has a stronger condensing action than dimethylcyclohexane.

An investigation of the alkylbenzenes (BOOIJ, VOGELSANG, and LYCKLAMA 1950) clearly showed that the action of these hydrocarbons on soap coacervates is the result of two opposing factors (Fig. 92). The long alkylbenzenes will get the same character as the paraffins. Thus, cetylbenzene has an opening activity, but this action is not so high as that of hexadecane. This fact is undoubtedly due the opposing effect of the benzene nucleus.

In an endeavour to elucidate the mechanism of plant growth hormones a number of aromatic

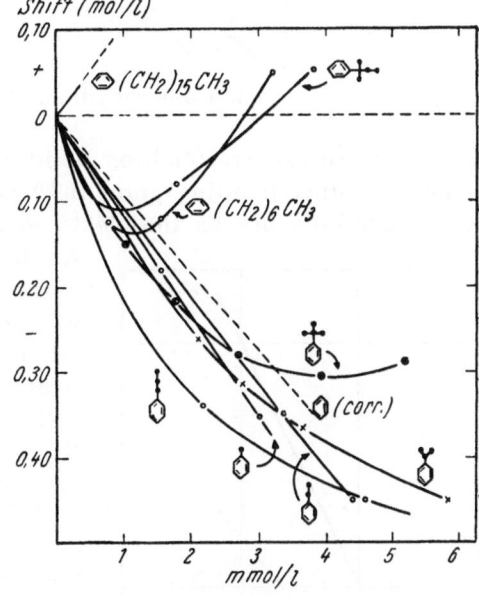

Fig. 92. The influence of an ever increasing alkyl group coupled to a benzene ring (oleate coacervate).

acids have been investigated as regards their action on the oleate coacervate (BOOIJ and VELDSTRA 1949, VELDSTRA 1952). It has not been possible to find a relation between the growth hormone activity and the action on the coacervate. However, the importance of the hydrophily/hydrophoby balance came clearly to the fore. Thus decahydronaphthalene acetic acid appeared to have a much higher activity than naphthalene acetic acid. This difference is in accordance to the general rule given above.

naphthalene acetic acid

decahydronaphthalene acetic acid

It was even possible to distinguish between the two isomers of 1, 2, 3, 4-tetrahydronaphthilydene acetic acid.

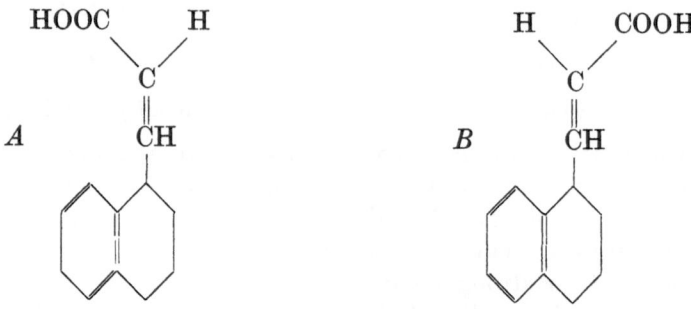

1, 2, 3, 4-tetrahydronaphthilydene acetic acid

One of these isomers (melting point 92⁰) is active as a plant growth hormone, the other (melting point 163⁰) is not. The higher melting isomer has a stronger influence on the oleate coacervate than the lower melting one.

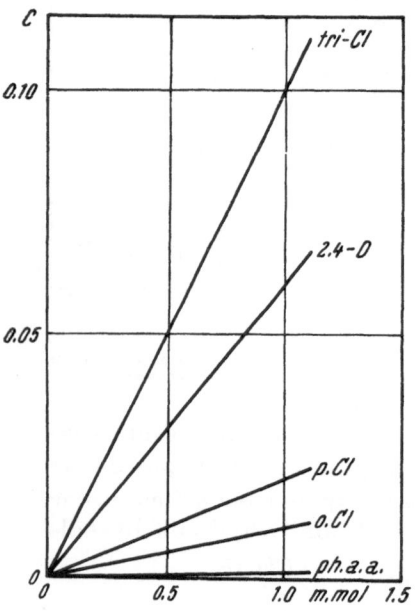

Fig. 93. Activity of chlorinated phenoxy-acetic acids (ph. a. a. = phenoxy-acetic acid, *o*-Cl = the orthochloro derivative, 2, 4-D = 2, 4-dichloro phenoxy-acetic acid, tri-Cl = 2, 4, 6-trichloro phenoxy-acetic acid).

As the total length of a molecule is of importance to its action on the coacervate we concluded that the higher melting isomer must have the configuration B (trans-compound). From its physiological activity it was deemed very possible that the lower melting substance has the cis-configuration (A).

When we compare these acids with the aliphatic ones we see e. g. that γ-naphthalene (2) butyric acid is about as active as decanoic acid, while naphthalene (2) acetic acid has the same activity as octanoic acid. From these experiments we may deduce that the "aliphatic equivalent" of the naphthalene nucleus is about 6 carbon atoms. The aliphatic equivalent of the decahydronaphthalene nucleus is much higher (about 8 carbon atoms). Experiments on phenylbutyric acid showed that the benzene nucleus has an aliphatic equivalent of approximately 3 carbon atoms.

An interesting example of the effect of the introduction of hydrophobic groups we found when studying phenoxyacetic acid and its chlorinated derivatives. The activity on the oleate coacervate increases strongly on chlorination (Fig. 93). The hormone activity increases up till the well-known 2, 4-D. Here, presumably, the proper balance between hydrophobic and hydrophilic groups in the molecule is reached. A further increase of the number of hydrophobic groups diminishes the growth activity considerably. Compare also naphthalene acetic acid (proper balance; growth hormone) with decahydronaphthalene acetic acid (too hydrophobic; no growth activity).

The investigation of chlorinated benzoic acids brought to light a new phenomenon (VELDSTRA 1952). This may be illustrated by the activity of the trichlorobenzoates (Fig. 94). By far the strongest activity on the oleate coacervate is found when the chloro-atoms are situated at the end

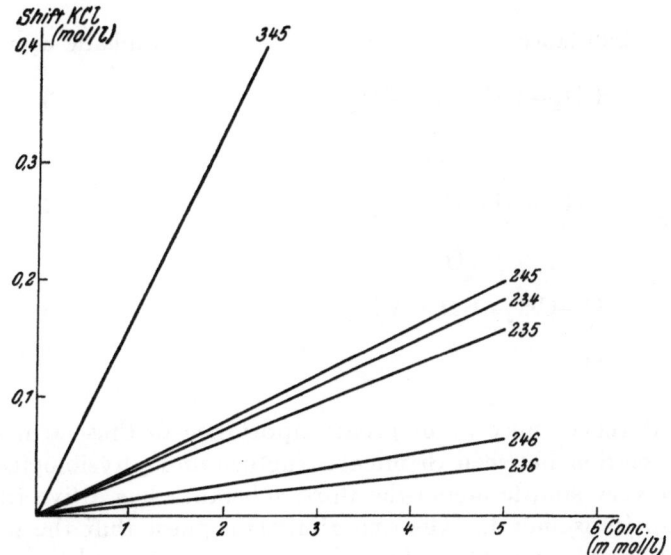

Fig. 94. Opening action of trichlorobenzoic acid anions on the oleate coacervate.

of the molecule, opposite to the carboxyl group. Ortho-substitution interferes with activity on the oleate (in parentheses: only 2, 3, 6-trichlorobenzoic acid does show plant growth activity). U. V.-absorption spectra showed that ortho-substitution (and especially di-ortho-substitution) interferes with the conjugation between the benzene nucleus and the carboxyl group. When interpreting the experiments with the oleate coacervate, we might conclude that a benzene nucleus in conjugation with a carboxyl group is much more hydrophobic than a benzene nucleus not in conjugation.

A study of the action of aromatic alcohols and ketones on the oleate coacervate provided new information on the question of the aliphatic equivalent of the benzene nucleus (BOOIJ, BLEKKINGH, and KWESTROO-VAN DEN BOS 1954). Phenylethylalcohol has about the same activity as n-amylalcohol. The same applies to the pair phenylpropylalcohol and n-hexylalcohol. Thus we may conclude that the aliphatic equivalent of the benzene ring in these substances is 3 carbon atoms. This seems to be the normal aliphatic equivalent of the benzene nucleus (see also page 96 for phenylbutyric acid).

The aromatic ketones showed the same value, provided that the benzene nucleus and the carbonyl group were separated by at least one methylene group. In ketones with the phenyl group next to the carbonyl group the value ist 5 carbon atoms. Thus it seems that by interaction with side chains (redistribution of electrons) the benzene nucleus may get a more hydrophobic character (compare the influence of conjugation in benzoic acid, quoted above). The introduction of a conjugated double bond between the polar group and the benzene nucleus results in a increase of the latter's aliphatic equivalent. An example of the values found is given below.

Substance	Aliphatic equivalent
I $\langle\!\!\!\!\bigcirc\!\!\!\!\rangle$—CH$_2$—CH$_2$—C(=O)—CH$_3$	± 3
II $\langle\!\!\!\!\bigcirc\!\!\!\!\rangle$—CH=CH—C(=O)—CH$_3$	± 4
III $\langle\!\!\!\!\bigcirc\!\!\!\!\rangle$—C(=O)—CH$_2$—CH$_2$—CH$_3$	± 5

These differences may be of great importance to those who are trying to find the relation between chemical structure and physiological activity. Even in our very simple model the third substance has a five times higher activity than its isomer I. The experiments suggest that the introduction of a benzyl group in a physiologically active compound may sometimes have less effect than the introduction of a phenyl group (though the former has a larger number of C-atoms).

In aromatic ethers we found the same phenomenon (Booij, Kwestroo-van den Bos and Blekkingh 1954). The aliphatic equivalent of the benzene ring is approximately five for phenylethers and three to four for benzylethers. Here too it was suggested that in the case of phenylethers the benzene nucleus has grown more hydrophobic, presumably by the deactivating influence of the ether-oxygen.

In general one might say that the possibilities of interactions between the side chains of a benzene nucleus are so many that it is difficult to predict the action of a substance containing a benzene ring (Booij 1952). From the studies with aliphatic branched molecules one would predict that the activity in a series of aromatic isomers containing a strongly hydrophilic group and a hydrophobic group increases according to the sequence $o < m < p$ (Fig. 95). From the left to the right the molecule grows longer. Thus this series might be compared to the aliphatic series: tert.-butylalcohol < sec.-butylalcohol < isobutylalcohol < n-butylalcohol.

When we introduce—beside a hydrophilic anchor group—a second hydrophilic group, we observe a pronounced fall in activity. We are inclined to suggest that the rule given in Fig. 95 is reversed in this case (see Fig. 96). It is of importance to note that some chloroderivatives of aromatic substances give the sequence pictured in Fig. 95, for other chloro-derivatives the reverse has been found. In the case of the latter substances (e. g. chloroanilines) we are dealing with molecules containing an electron donor as anchor group, while the second side chain is electron attracting (Fig. 97). Then this second group gets a stronger hydrophilic character than it had originally.

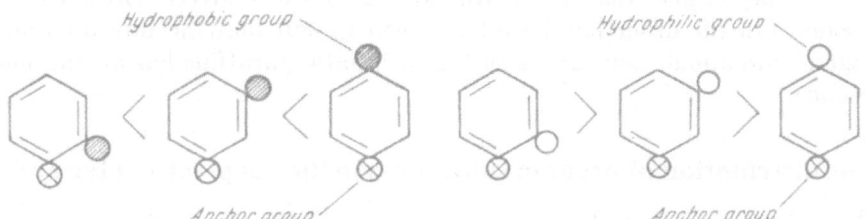

Fig. 95. Sequence of the action of benzene deri-vatives with a (strongly) hydrophilic anchor group and a hydrophobic second substituent.

Fig. 96. The sequence of Fig. 95 is reversed if the second group has a relatively hydrophilic character.

It is still difficult to connect the experimental results with known data of theoretical organic chemistry. Many effects will, presumably, have an influence on the activity of aromatic substances. If we limit ourselves to aromatic compounds with one hydrophilic anchor group and a second sub-stituent we might mention some possibilities:

a) When there is no interaction between the second group and the rest of the molecule we will find the series $o < m < p$, if the second group is a hydrophobic one. Should the second group be hydrophilic, then a reverse sequence will be expected.

b) In most cases there will be more or less interaction between the second group and the rest of the molecule. The second group may assume a relatively hydrophilic character if it is electron attracting, while the anchor group is an electron donor (sequence $o > m > p$).

Fig. 97. If the anchor group is an electron donor, while the second group is electron attracting, the latter group may get a relatively hydrophilic char-acter.

c) It is to be expected that situation b) will arise only if the respective groups are strongly electron attracting or releasing. If not, resonance effects may produce shifts in the sequences. It has been found, e. g., that m-chlorophenol has a stronger activity than the o- and p-isomers.

These kind of experiments may in future shed some light upon pharma-cological phenomena. It is sometimes said that the introduction of a chlorine atom into a molecule has—pharmacologically speaking—the same influence as the introduction of a methyl group. This "rule" is far from a general one. It might be supposed that the chlorine-methylgroup ana-logy must be understood in the following way. If we introduce into a molecule a methylgroup or a chlorine atom, then the resulting molecules will show—in many cases—approximately the same distribution in the living cell. However, the chlorine atom tends much more to interaction with other groups in the molecule (especially via aromatic rings) than the methyl group. Thus the "rule" quoted before depends very much on the question into which molecule the chlorine atom is introduced. From our experiments on chloro-paraffins and aromatic chloro-derivatives we are inclined to suggest that the rule is only valid when: a) there is no inter-action (or only slight interaction) between the chlorine atom and the rest of the molecule and b) the molecule contains another hydrophilic group. Thus we may expect that many aromatic chloro-derivatives will not follow the rule. On the other hand we have seen (p. 89) that the introduction of halogens into apolar substances such as aliphatic paraffins has an enormous influence.

e) The distribution of organic substances in the soap/salt/water system

In order to explain the experiments on soap coacervates we suggested that added organic substances are distributed among several positions, pictured in Fig. 81. Some substances are found preferentially in the medium, other molecules will concentrate in the micelles. We will now

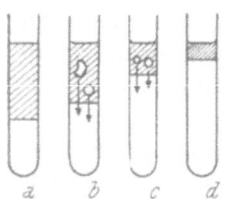

ask whether it is possible to get more information on this hypothetical distribution. As regards the influence of organic substances on soap coacervates especially the positions *a*, *c* and *d* of Fig. 81 are of importance. First of all we will pay attention to the distribution medium/micelle.

An experiment on the influence of *o*-cresol on the oleate coacervate demonstrates our view on the distribution equilibrium medium/micelle clearly. To 50 ml. of an oleate solution (2%; extra KOH added) we weighed 0.75 g. *o*-cresol. With the aid of KCl we made a coacervate of this soap solution. A coacervate layer was formed in a relatively short time, but it

Fig. 98. An oleate coacervate with added *o*-cresol is not stable. Its volume decreases till—after several hours—a stable, much more condensed, layer has been formed.

was not stable (Fig. 98). Large vacuoles were formed which united with the equilibrium liquid. The next day the coacervate layer was very much smaller (*d*). In normal cases, e. g. when working with alcohols, paraffins, ketones etc., this phenomenon never occurs. In the case of *o*-cresol we are

dealing with a substance, which—at the high pH of these oleate solutions—is dissociated considerably. The molecule has a condensing action, the anion will show practically no activity (compare the fatty acid anions). When adding a certain amount of KCl a coacervate will be formed. The micelles then contain a small amount of undissociated cresol molecules. This preferential uptake of molecules causes a shift in the dissociation equilibrium. More molecules are formed, which are again taken up by the micelles. Thus the coacervate will grow ever more condensed till a new equilibrium is reached. Additionally, this experiment gives a strong argument for the hypothesis that the micelles in the coacervate are otherwise than those in a soap solution. Indeed, when the micelles were of the same structure the taking up of undissociated molecules would already have occured. Then the phenomenon pictured in Fig. 98 would not take place.

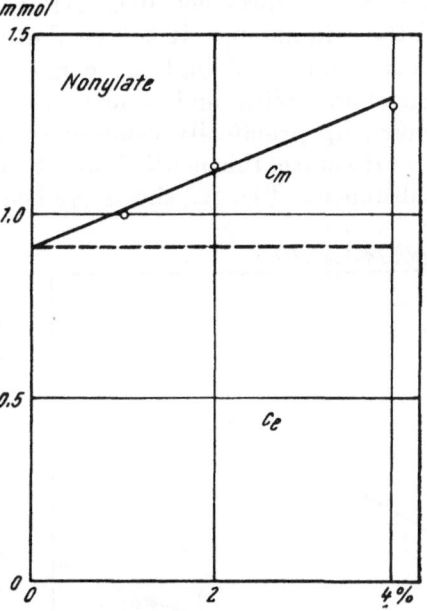

Fig. 99. The quantity of nonanoate required to cause a shifting of the KCl curve of 0.04 mol/l. (oleate coacervate). We find an equilibrium concentration of 0.91 m mol nonanoate.

When our hypothesis as regards the distribution equilibrium medium/micelle is right we would expect that the total concentration of any organic substance consists of two parts: molecules in the medium and molecules taken up in the micelles. When augmenting the soap concentration we need (generally speaking) more added organic substance to get the same degree of coacervation. This is to be expected as the number of micelles increases and more substance can be taken up. On the other hand the concentration in the medium will not be altered (at least approximately, as the soap-concentration will always be rather low as compared to the amount of water present). Thus experi-

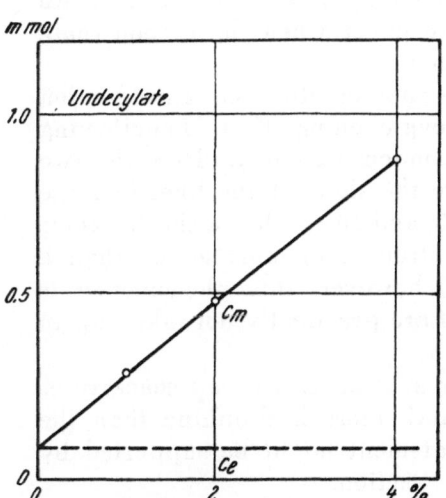

Fig. 100. For undecanoate the equilibrium concentration is much lower than for nonanoate (compare Fig. 99).

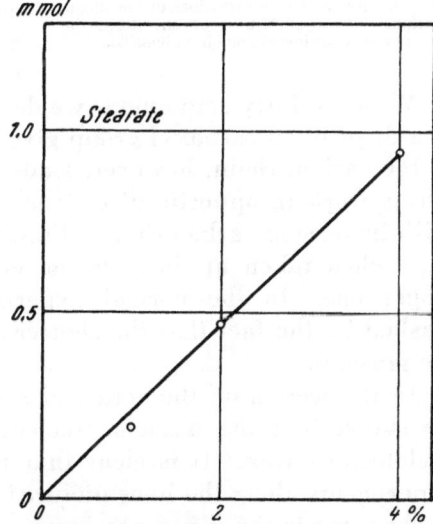

Fig. 101. For stearate the equilibrium concentration drops to zero (compare Figs. 99 and 100).

ments on the action of organic substances at various soap concentrations will give us some idea about the distribution of those substances. The following figures (99, 100, 101) show the results of experiments with some fatty acid anions. It is seen that in the case of nonanoate the equilibrium concentration is high as compared with the amount taken up in the micelles. With undecanoate the situation is reversed, while stearate is taken up practically completely in the micelles.

Of course, this method may be used with opening as well as condensing substances. Fig. 102 shows the behaviour of normal alcohols. Here too the equilibrium concentration decreases with increasing chain length, to become practically zero for *n*-octylalcohol. It is noteworthy that the experiments with alcohols and fatty acid anions show a difference. With the fatty acid anions the slope of the line increases with increasing length of the carbon chain. With the normal alcohols the reverse takes place. This difference would be expected from the hypothesis on the mechanism of action [24]. We pictured the large flat micelles as a result of two opposing forces. In the case of the alcohols the uncharged group will give a condensing action. Thus both effects work in the same direction. This results in a decreasing steepness of the lines: of the molecules taken up less is needed to reach a certain effect when the carbon chain is lengthened.

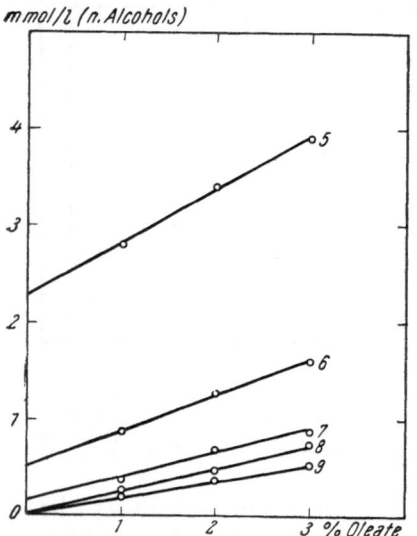

Fig. 102. The quantity of normal alcohols needed to shift the KCl curve a certain amount varies for different concentrations of the oleate solution. The equilibrium concentrations decrease with increasing chain lengths.

With the fatty acid anions we deal with another situation. Introduction of a dissociated carboxyl group gives a strong opening effect. Lengthening of the carbon chain, however, leads to a condensing effect. Here the two effects work in opposite directions. Hence the slope of the lines increase with increasing carbon chains (Figs. 99, 100, and 101). Thus a shorter anion has—when taken up into the micelle—a stronger opening action than a longer one. In the normal experiments, however, this phenomenon is masked by the fact that the shorter anions are practically not taken up in the micelles.

In the section on the action of aromatic substances on soap coacervates we stated that the benzene nucleus is much more hydrophilic than the cyclohexane ring. It is clear that this statement might be supported by experiments along the lines pictured in this section.

As regards the difference between benzene and cyclohexane (the latter

[24] See p. 83.

having a stronger condensing action) we expect that the equilibrium concentration of the former will be higher than that of the latter. Fig. 103 (Booij, Vogelsang, and Lycklama 1950) shows that the experiments are in accordance to the expectation. The slope of the two lines is practically equal. This means that the activity of benzene and cyclohexane, when taken up in the micelle, is comparable. The difference in action in our experiments must be ascribed solely to the difference in distribution equilibrium medium/water.

The same applies to the difference in action between benzylalcohol and cyclohexylcarbinol (Fig. 104; Booij, Blekkingh, and Kwestroo-van den Bos 1954). Here too we observe a large difference in equilibrium concentration, but the slopes of the lines seem to be approximately equal.

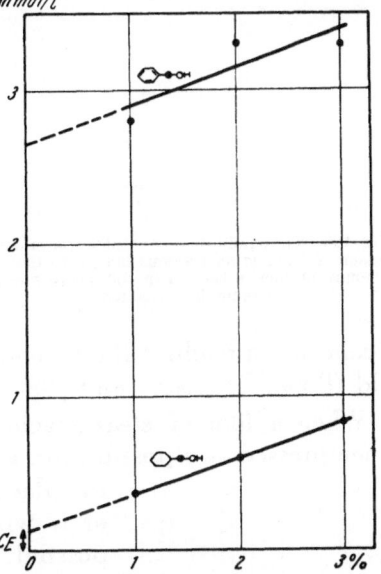

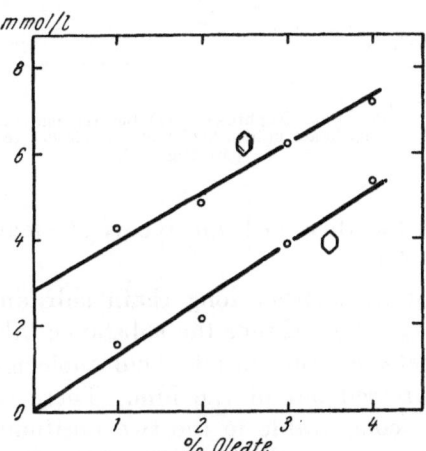

Fig. 103. Influence of benzene and cyclohexane at different oleate concentrations.

Fig. 104. The difference in action of benzylalcohol and cyclohexylcarbinol on the oleate coacervate must be ascribed to the large difference in equilibrium concentration.

We have seen that the plant growth substance naphthalene (1) acetic acid has only a small action on the oleate coacervate. When investigating this compound in various soap concentrations it was found indeed that the equilibrium concentration is large (Fig. 105). The amount of molecules taken up into the micelles is so small that it cannot be detected with this method. Increasing the carbon chain changes the picture (Fig. 106). Here an appreciable number of molecules enter the micelle.

It is more difficult to get some information about the distribution of organic molecules within the micelles (see Fig. 81). Three ways of approach might be mentioned here.

First of all we tried to get some support for the hypothesis that long apolar molecules would be concentrated in between the CH_3-planes of the micelles. We tried to get this additional evidence from experiments on the spreading of molecules of this character on the interface water/air. It is a well known fact that it is not possible to measure the force/area diagram of long apolar molecules with weakly polar groups (e. g. octadecylbromide or octadecylbenzene). Either no film is formed or the film collapses at very low pressures. We have tried to imitate the conditions in the soap

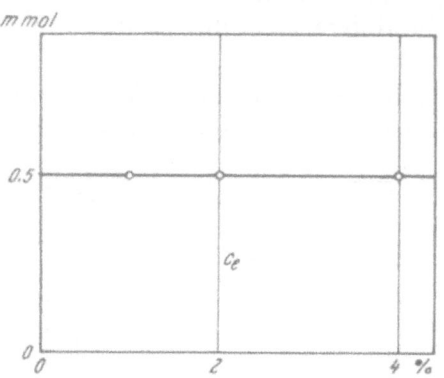

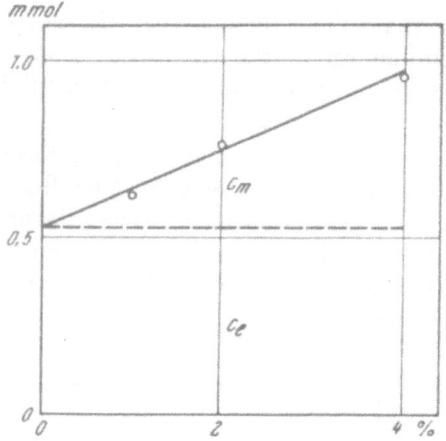

Fig. 105.   The equilibrium concentration of naph-thalene (1) acetic acid anions is so high that the amount of ions taken up in the oleate micelles cannot be measured.

Fig. 106.   Naphthalene (2) butyric acid shows a moderate uptake in the oleate micelles (com-pare Fig. 105).

micelle by spreading these compounds together with an excess of stearic acid (Booij, Lycklama, and Vogelsang 1950).

When a film of stearic acid containing another long chain substance is compressed two possibilities arise (Fig. 107). Either the substance takes

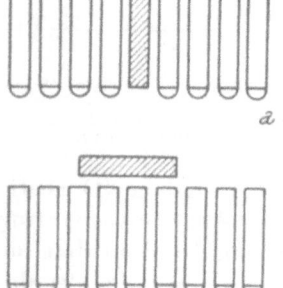

Fig. 107. Possibilities of spread-ing an apolar substance with a weakly polar group in an ex-cess of stearic acid.

its place in between the stearic acid molecules, or it will be pressed out of the film. These two possibilities are comparable to the two positions c and d in the soap micelle (see Fig. 81). When spreading stearic acid and octadecylalcohol to-gether the area (computed for a stearic acid mole-cule alone) appears to be too large (Fig. 108). Thus the long alcohol remains parallel to the stearic acid in the compressed film. On the other hand, when spreading stearic acid and hexadecane to-gether the force/area diagramm does not change. We must conclude that hexadecane is easily pressed out of the film. In terms of the oleate micelle: hexadecane prefers position d, while octadecyl-alcohol will concentrate at c (Fig. 81). We saw that the introduction of a weakly polar group into a long paraffin resulted in a large change in character (see Fig. 84). Spreading experiments with octadecylbenzene (Fig. 109) were in agreement with these experiments on soap coacervates. The benzene nucleus proved to be a—be it weak—anchor to the water surface. The difference with octodecylalcohol may be judged from the fact that films containing more than 20% octadecylbenzene collapse, while stearic acid/octadecylalcohol films are formed at any proportion of the two compounds.

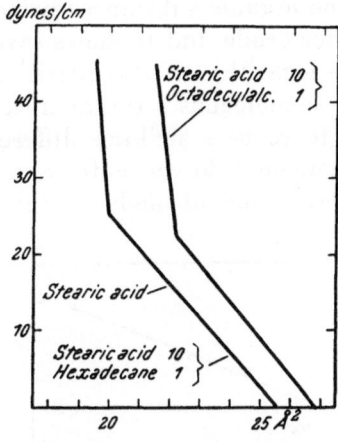

Fig. 108. Force/area diagrams of a) stearic acid, b) a mixture of stearic acid and octadecylalcohol and c) a mixture of stearic acid and hexadecane.

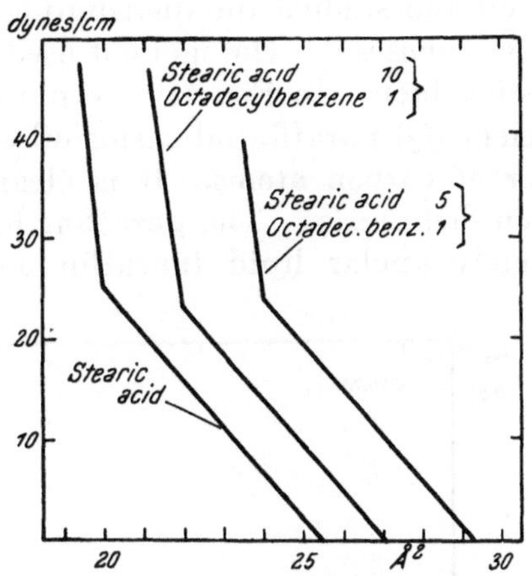

Fig. 109. Force/area diagrams of mixtures of stearic acid and octadecyl benzene.

The second approach has been a study of the distribution of organic substances between two liquids of lipid character. One of these liquids

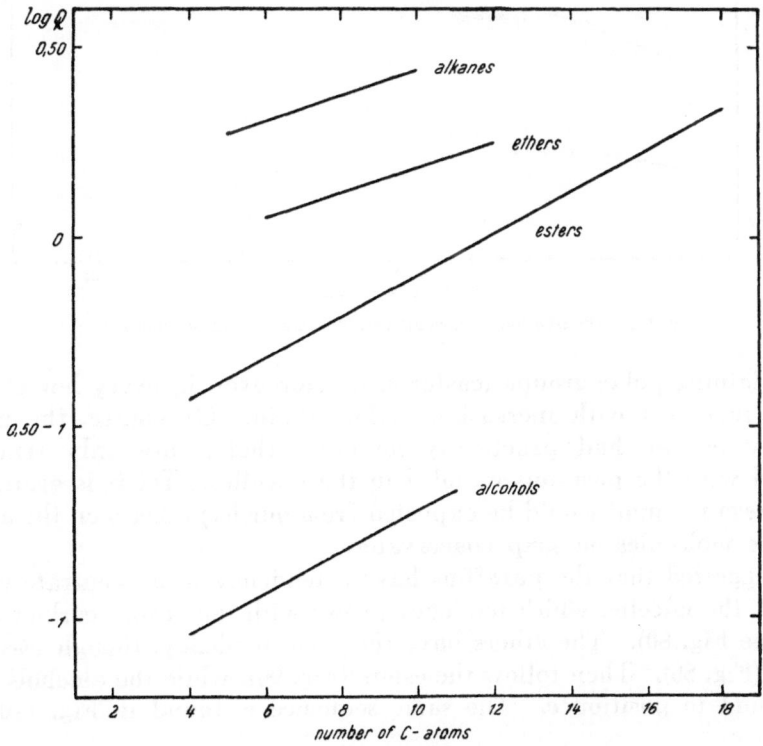

Fig. 110. Distribution coefficients paraffin oil/castor oil of some homologous series.

must have the character of position d (Fig. 81), the other that of position c. Moreover the two liquids must not be miscible. We used paraffin oil and castor oil and studied the distribution of some organic substances between these two phases [25]. The method used is rather crude, but it shows several interesting facts. In Fig. 110 we plotted the logarithm of the distribution coefficient (Q) paraffin oil/castor oil of some homologous series against the number of carbon atoms. It is clear that there is a striking difference between these series. The paraffins have a tendency to concentrate in the completely apolar lipid (paraffin oil), whereas the alcohols prefer the

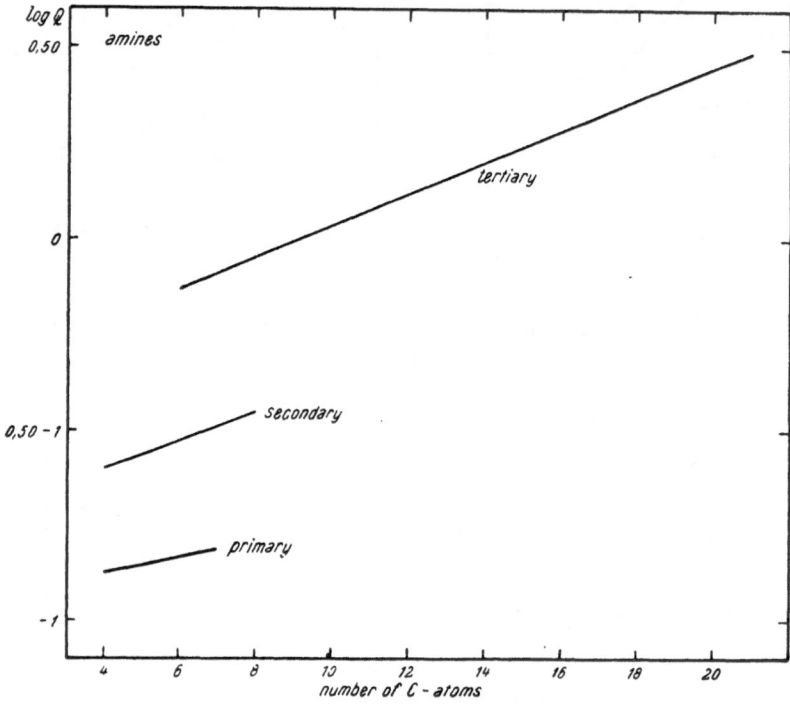

Fig. 111. Distribution coefficients paraffin oil/castor oil of amines.

lipid containing polar groups (castor oil). Moreover, in every homologous series Q increases with increasing carbon chain. Of course, the media chosen by us—we had practically no other choice—are only remotely connected with the positions c and d in the micelles. Yet it is gratifying that the results found would be expected from our hypothesis on the action of organic molecules on soap coacervates.

We suggested that the paraffins have a tendency to concentrate in the middle of the micelle, which tendency grows with increasing carbon chain length (see Fig. 80). The ethers have the same tendency, though less pronounced (Fig. 86). Then follow the esters (Fig. 90), while the alcohols seem to be bound to position c. The same sequence is found in Fig. 110 and

---

[25] Booij, not yet published.

the same effect of increasing chain length comes to the fore. It goes without speaking that we may only expect a qualitative agreement here.

A study of some amines gives another interesting phenomenon (Fig. 111). The primary amines resemble the normal alcohols very much (as regards the oleate coacervates this has been found already by ROSENTHAL in 1939). When the polar group is screened off by alkyl groups the amines (secondary and specially tertiary) are taken up much more readily by the apolar liquid.

Finally we may quote some experiments by BOOIJ and VAN MULLEM (1951). They reasoned that it would be possible to influence the distribution equilibrium of an organic substance within the micelle by adding a second compound. If e. g. a substance is distributed equally among the places c and d (Fig. 81), then this equilibrium would be disturbed by adding a substance known to concentrate at c. For the second substance we chose amylalcohol. This alcohol will concentrate parallel to the oleate molecules. As its chain is much shorter than that of the soap molecules, every amylalcohol molecule taken up in the micelle will create a kind of "free space" extending from the end of the molecule to the middle of the micelle [26]. This "space" might be filled up by an apolar substance originally present in the middle section of the micelle (so this substance might move from position d to c with a simultaneous change of action). This seems to be a more favourable situation from an energetic point of view than the disorder originally present.

The experiments were in line with the hypotheses framed. We have seen (Fig. 80) that the paraffins have an opening action, which comes clearly to the fore at heptane and increases considerably when increasing the chain length. When amylalcohol is added to the soap system we find that the opening action of all paraffins from hexadecane down till hexane decreases. Nonane e. g., which has normally an opening action at low concentrations, behaves like a condensing substance when 44 m mol/l. amylalcohol is present in the system. The condensing action of hexane is practically not disturbed by the addition of amylalcohol, while pentane gets a less condensing activity. The experiments may be summarised in the following scheme.

Number of carbon atoms of the alkanes: 5 6 7 8 9 10 11 etc.
Opening action after
    addition of amylalcohol: ↑   —   ↓   ↓   ↓   ↓   ↓
(↑ = opening action of alkane increased, ↓ = opening action decreased.)

This tallies well with our hypothesis on the distribution of organic apolar substances in the soap micelles. The condensing action of pentane has been attributed to the fact that this substance would be concentrated more or less at position c. Then amylalcohol would shift the equilibrium and pentane would get a less condensing activity. The higher paraffins would show a reverse situation, while hexane seems to take the intermediate position.

---

[26] Compare also page 86 (small print).

We remind of the strong influence which the introduction of a double bond into a paraffin has on its action (Fig. 82). The rather strong condensing action which results from the introduction of a double bond has been attributed to its relatively hydrophilic character. Of course, amylalcohol can compete successfully with substances like 3-methylheptene-2, transoctene-4 and octyn-4 for position c in the micelle. Thus addition of amylalcohol diminished the condensing action of these compounds strongly (while it has the reverse influence on methylheptane). Fig. 112 demonstrates the effect of added amylalcohol on the activity of n-hexylbromide. We see that in this case too, the condensing action decreases.

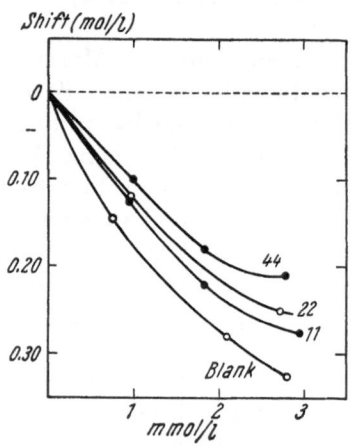

Fig. 112. The influence of amylalcohol on the action of n-hexylbromide.

There exists a distinct difference between naphthalene and decahydronaphthalene as regards their action on an oleate coacervate (the former substance having a stronger condensing action). This difference in action has also been ascribed to a difference in distribution in the micelle (Booij 1949). Thus it was not astonishing to find that the addition of amylalcohol effects the action of the two compounds in an opposite way. Naphthalene becomes a substance with less pronounced condensing activity, while decahydronaphthalene gets a stronger condensing action. The addition of amylalcohol tends to diminish the difference between the two substances.

f) The influence of alcohols on the fixation of ions

The shorter members of several homologous series (alcohols, ketones etc.) show an opening action on soap coacervates. This phenomenon is not readily understandable from the hypothesis given in the preceding pages. As we have seen the phenomenon depends strongly on the "degree of condensation" in the soap micelles. An alcohol having a condensing action on a certain coacervate may have an opening action if this coacervate is more condensed previously. This would lead one to the supposition that the shorter terms may act at the surface of the micelles, especially in those cases where the soap micelles are strongly condensed.

In the case of oleate coacervates this supposed action on the surface of the micelles will not be easily demonstrable. We might suppose that some interaction with the cations would occur, but the amount of salt needed to get coacervation is so high, that the amount of cations attached to the micelles is only a very small fraction. Variations in this amount under the influence of alcohol would not be measurable.

The discovery of the coacervation of cetyltrimethylammoniumbromide

(CTAB) with KCNS brought new possibilities. Here the amount of anions needed for coacervation is low and it may be expected that differences in the amount attached to the micelles can be detected.

The coacervate obtained from CTAB by the addition of KCNS is influenced by alcohols in the way already discussed (BUNGENBERG DE JONG and RECOURT 1953). The shorter alcohols show an opening (salt-demanding) action, the longer ones (from amylalcohol upwards) condense the coacervate. When working with different CTAB concentrations (compare p. 52) it is possible to determine the influence of alcohols on the binding of anions. In Fig. 113 the results of this investigation are given. It is seen that ethanol shifts the blank curve upwards with a nearly unaltered slope. This means that the amount of anions bound to the micelles (degree of occupation) required for coacervation is not altered by ethanol (a concentration of 1.03 mol/l. has been used). On the other hand the equilibrium concentration of KCNS required to reach this degree of occupation is increased greatly (20.9 millimoles/l. to 47.7 millimoles/l.). This must be due to a weakening of the fixation of CNS-anions. In this case the salt-demanding (opening) action of a short alcohol is indeed an influence exerted at the surface of the micelles. Though it cannot be proved, it seems not unreasonable to assume that in other soap coacervates too the opening action of short terms of homologous series must be ascribed to a weakening of the ion fixation to the micelles.

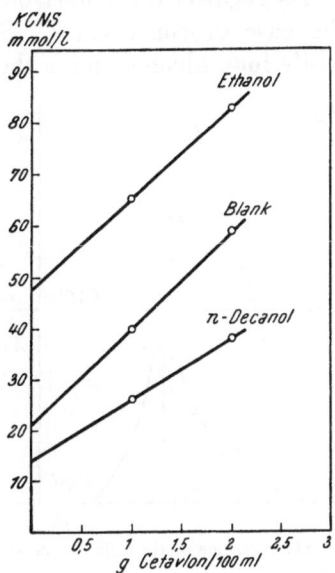

Fig. 113. Influence of ethanol and n-decanol on the coacervation of CTAB (cetavlon) as a function of the soap concentration.

On the other hand the reversal of charge concentrations of cations is strongly decreased after the addition of alcohol, acetone etc., when working with macromolecules as e. g. gum arabic. There we are inclined to think that the affinity between the cations and the negative groups of the colloid is increased. Presumably the influence of the alcohol on the solubility of the association colloid plays a decisive role.

The influence of n-decanol on the CTAB coacervate is quite different. Here—in accordance to the hypothesis given before—less anions are required to reach the same coacervation. This is found in two ways. The n-decanol line has a smaller slope (smaller degree of occupation at the coacervation limit) and the point of intersection with the ordinate lies lower (lower equilibrium concentration).

g) The influence of organic substances on elastic-viscous systems

As we have assumed that the structural units of elastic-viscous systems and coacervates of lipophilic character are the same, we must expect that the influence of organic substances on the former systems is fundamentally comparable to that on the latter. Several substances have been investigated as regards their influence on the oleate elastic-viscous system. In the scheme of Fig. 114 the results have been summarised. As regards the shifts in horizontal directions three possibilities occur. An organic substance may shift the inflexion point of the G-curve or the maximum of the $1/\varDelta$, λ or n curves to the left [27]. This means a salt-sparing influence. A shift to the right is of course, a salt-demanding activity.

As regards these horizontal shifts we find the same rules which hold in the case of coacervates. Short alcohols have a salt-demanding activity, while long alcohols have the opposite effect. Also in the homologous series

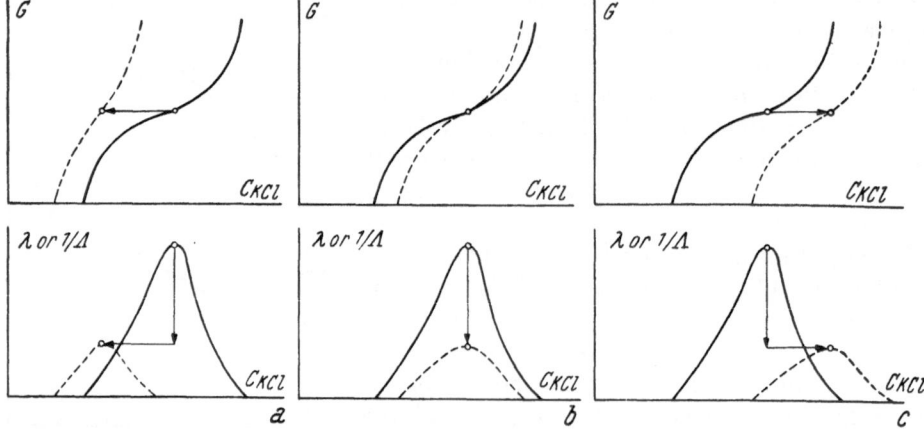

Fig. 114. Scheme of the influence of organic substances on the elastic behaviour at constant oleate concentration (full line = blank).

of the alkanes and alkylbenzenes a transition from salt-sparing to salt-demanding activity comes to the fore (see references of chapter 6: VI, VII, VIII, XI, XII and XVIII).

From Fig. 114 we see, however, that there exists another effect, which is always in the same direction. The maximum in the $1/\varDelta$ or $1/n$ curve is lowered under the influence of the added organic substance, be it a salt-sparing, a salt-demanding compound or a substance in which these two tendencies balance. This means that the damping always increases; the micelles get a more supple character (compare p. 73).

h) Biological perspectives

One would be inclined to suppose that the condensing influence of a certain substance on the oleate coacervate might mean that this substance would also have a condensing influence on the protoplasmic membrane, in other words decrease permeability (see also chapter 9). This is not necessary. First of all the protoplasmic membrane may be a much more condensed lipid structure. Secondly the experiments on the elastic-viscous

---

[27] Compare page 68.

systems showed that the micelles grow more supple under the influence of all kinds of organic substances. Translated into terms of permeability this might mean that all of these substances make the micelle more permeable.

As ever more evidence is gathered from histological studies on the living cell (electronoptics) that double layers of lipids are found everywhere within the cell, the importance of the experiments on synthetic double layers of lipids is obvious. In our opinion the most important finding is that beside the distribution medium/micelle another equilibrium (within the micelle) plays an important rôle. These two equilibria might also occur in biological double layers of lipids. The action of organic substances is largely dependent on the two distribution coefficients involved. It is regrettable that experiments along the lines developed in this chapter on phosphatids and other lipids of biological importance are not yet feasible. We hope that the experimental difficulties in this field will soon be over-come, as the solution of many biological problems would benefit from this kind of research.

One experiment on lecithin might be quoted. We have already seen that salicylate at high concentration may cause the coacervation of a lecithin suspension (here salicylate has an opening action; we start at the smectic phase; see p. 60). A number of short alcohols (octanol included) have an opening action on this lecithin/salicylate coacervate. Only at nonanol we reach the first condensing alcohol. In the light of our discussion this must mean that the lecithin micelles are extremely condensed (which fact is not surprising as the molecule contains two long carbon chains). The theoretical importance of this find for biological lipophilic systems is obvious: we must indeed suppose that the phosphatide containing proto-plasmic membrane is strongly condensed.

References

Booij, H. L., 1949: Chapter XIV in Kruyt's Colloid Science II. Amsterdam.
— 1949: Influence of organic compounds on soap and phosphatide coacervates IX. Proc. Kon. Ned. Akad. Wetensch. Amst. 52, 1100—1110.
— 1952: Influence of organic compounds on soap and phosphatide coacervates XVII. Rec. trav. chim. Pays-Bas 71, 101—107.
— 1952: Neuere Untersuchungen über Koazervate von Seifen. Kolloid-Z. 125, 21—31.
— J. H. Blekkingh, and H. Kwestroo-van den Bos, 1954: Influence of organic compounds on soap and phosphatide coacervates XXI. Proc. Kon. Ned. Akad. Wetensch. Amst. B 57, 340—350.
— and H. G. Bungenberg de Jong, 1949: Influence of organic compounds on soap and phosphatide coacervates VIII. Biochim. Bioph. Acta 3, 242—259.
— and E. S. van Calcar, 1950: Influence of organic compounds on soap and phosphatide coacervates XIV. Proc. Kon. Ned. Akad. Wetensch. Amst. 53, 1169—1177.
— H. Kwestroo-van den Bos, and J. H. Blekkingh, 1954: Influence of organic compounds on soap and phosphatide coacervates XX. Proc. Kon. Ned. Akad. Wetensch. Amst. B 57, 215—226.
— and A. M. van Leeuwen, 1953: Influence of organic compounds on soap and phosphatide coacervates XVIII. Proc. Kon. Ned. Akad. Wetensch. Amst. B 56, 255—267.
— J. C. Lycklama, and C. J. Vogelsang, 1949: A study on the refraction of coacervates. Proc. Kon. Ned. Akad. Wetensch. Amst. 52, 1006—1016.

Booⅈ, H. L., J. C. Lycklama, and C. J. Vogelsang, 1950: Influence of organic compounds on soap and phosphatide coacervates XII, Proc. Kon. Ned. Akad. Wetensch. Amst. 53, 407—413.
— — — 1950: Influence of organic compounds on soap and phosphatide coacervates XV. Proc. Kon. Ned. Akad. Wetensch. Amst. 53, 1413—1421.
— and P. J. van Mullem, 1951: Influence of organic compounds on soap and phosphatide coacervates XVI. Proc. Kon. Ned. Akad. Wetensch. Amst. B 54, 273—290.
— and H. Veldstra, 1949: The effect of plant growth substances on coacervates. Biochim. Bioph. Acta 3, 260—277.
— C. J. Vogelsang, and J. C. Lycklama, 1950: Influence of organic compounds on soap and phosphatide coacervates X. Proc. Kon. Ned. Akad. Wetensch. Amst. 53, 59—69.
— — — 1950: Influence of organic compounds on soap and phosphatide coacervates XIII. Proc. Kon. Ned. Akad. Wetensch. Amst. 53, 882—890.
— and D. Vreugdenhil, 1950: Influence of organic compounds on soap and phosphatide coacervates XI. Proc. Kon. Ned. Akad. Wetensch. Amst. 53, 299—304.
Bungenberg de Jong, H. G., and H. J. van den Berg, 1950: Elastic viscous oleate systems containing KCl VII. Proc. Kon. Ned. Akad. Wetensch. Amst. 53, 7—18.
— H. L. Booⅈ und G. G. P. Saubert, 1937: Der Einfluß organischer Nichtelectrolyte auf Oleat- und Phosphatidkoazervate II, Protoplasma 28, 543—561.
— — — 1938: Der Einfluß organischer Nichtelectrolyte auf Oleat- und Phosphatidkoazervate IV. Protoplasma 29, 536—551.
— — — 1938: Der Einfluß organischer Nichtelectrolyte auf Oleat- und Phosphatidkoazervate VI. Protoplasma 30, 53—69.
— and A. Recourt, 1953: Influence of organic compounds on soap and phosphatide coacervates XIX. Proc. Kon. Ned. Akad. Wetensch. Amst. B 56, 451—466.
— und G. G. P. Saubert, 1937: Der Einfluß organischer Nichtelectrolyte auf Oleat- und Phosphatidkoazervate I. Protoplasma 28, 498—515.
— — und H. L. Booⅈ, 1938: Der Einfluß organischer Nichtelectrolyte auf Oleat- und Phosphatidkoazervate III. Protoplasma 29, 481—497.
— — — 1938: Der Einfluß organischer Nichtelectrolyte auf Oleat- und Phosphatidkoazervate V. Protoplasma 30, 1—38.
Koets, P., und H. G. Bungenberg de Jong, 1938: Der Einfluß organischer Nichtelectrolyte auf Oleat- und Phosphatidkoazervate VII. Protoplasma 30, 206—215.
Rosenthal, S., 1939: Invloed van organische niet-electrolyten op zeepcoacervaten. Thesis, Leiden.
Veldstra, H., 1952: Researches on plant growth regulators XXI. Rec. trav. chim. Pays-Bas 71, 15—32.
— and H. L. Booⅈ, 1949: Researches on plant growth regulators XVII. Biochim. Bioph. Acta 3, 278—312.

8. Interaction between Long Chain Electrolytes and Proteins

a) Introduction. The coacervation gelatin + detergent + salt

Among the conjugated proteins we find the lipoproteins, but the nature of this class of proteins is still somewhat mysterious. As regards the binding between protein and lipid several opinions have been brought to the fore, varying between the covalent bond to "physical" association. One of the most conspicuous properties of the lipoproteins is the fact that one cannot extract the lipids with the aid of ether, while extraction does succeed with a mixture of alcohol and ether. Thus it seems that alcohol loosens the binding between lipid and protein. This fact speaks in favour of a physical association between the components of the lipoproteins.

Bungenberg de Jong and Booⅈ-Van Staveren (1942) found that potassium salts bring about coacervation in a mixture of gelatin and oleate at high pH (9.2). The coacervate obtained in this way contains oleate as well as gelatin. This type of coacervation excites our interest, as it may be inhibited by alcohols. An existing coacervate may disappear after the addition of alcohols. The amount of alcohol needed for this phenomenon decreases strongly at increasing length of the carbon chain. Previous investigations had shown that coacervates formed from oleate alone disappear only after addition of the shortest members of the homologous series (methyl-, ethyl-, propyl- and butylalcohol at pH = 9.2), while amylalcohol and the higher alcohols promote coacervation. Here lies a difference between the simple oleate coacervate and the gelatin/oleate coacervate. The latter also disappears after the addition of higher alcohols (amyl- up till octylalcohol have been investigated). The two coacervates also differ in another respect. The KCl-concentration required for coacervation is much less for the gelatin/oleate mixture than for oleate alone. These differences between the two coacervates suggest that the gelatin is not a neutral admixture in the "oleate coacervate," but that the gelatin shows a marked interaction with the oleate. In other words, we are dealing—in a sense—with a model for lipoproteins.

After the war we became aquainted with the investigations of PANKHURST (1949) on the coacervation of mixtures of gelatin and alkylsulfates or other anionic or cationic detergents. The experimental conditions for obtaining these coacervates resemble those of our experiments very much (pH higher than the i. e. p. of gelatin; some salt must be present). Thus it was thought worth while to investigate whether a gelatin/alkylsulfate coacervate is also sensitive to alcohols. Using a technical preparation (a mixture of secondary alkylsulfates called "Teepol") we obtained coacervation with gelatin and salts (BUNGENBERG DE JONG and VAN LEEUWEN 1952). This coacervation could be suppressed indeed by the first four terms of the homologous series of normal primary alcohols. On Teepol alone all alcohols—including the first four terms—act in the opposite direction: they promote coacervation. Here too we find the same difference between coacervate with and without gelatin: a number of alcohols promoting coacervation of the detergent alone, suppress coacervation of the combination detergent + protein. Furthermore, in this case too, the salt ($MgCl_2$) concentration needed to obtain coacervation is much smaller than that needed in the case of the detergent alone. Many salts which do not even coacervate the detergent alone (KCl, NaCl, LiCl, KNO_3, KCNS, K J, KBr) show coacervation of the combination Teepol + gelatin.

These facts strongly suggest an interaction between the gelatin and the detergent. Of course, the distance between the natural lipoproteins and our models is still large, as gelatin is not a globular, but a linear protein (we have the impression that many lipoproteins contain globular proteins), while the natural lipids are not detergents. Still the study of our gelatin/detergent models is interesting as it may be possible that some idea about the binding of the components emerges. Moreover, optical studies on living material have long ago led to the opinion that associations of linear proteins and lipids play a role in protoplasm.

b) The binding between protein and detergent according to the hypothesis of PANKHURST

The studies of PANKHURST on the association of alkylsulfates (and other detergents) and gelatin showed that there exist two limits for the ratio detergent/gelatin in the association. The amount of detergent bound to gelatin may vary between the following extremes: a) the amount equivalent to the number of oppositely charged groups of the gelatin and b) an amount which is about ten times as high. The transition $a \to b$ is promoted by increase of pH and increase of salt concentration. Finally—at sufficiently high salt concentration—a maximal value is reached, which does not increase at further addition of salt and which is independent of the salt used. As this maximal value agrees more or less with the number of

Table 5 (according to PANKHURST).

Amphipathic ion	Cross section		Estimated penetrability into the backbone (in %)
	$Å^2$ chain	$Å^2$ head	
Primary alkylsulfate	20	25[1]	100
Primary alkylamine	20	20	100
7-Alkylnaphthalene-3-sulfonate	35	25[1]	62
Secondary alkylsulfate	40	25[1]	46
Alkyl trimethylammonium	20	49 (31)	0

[1] These values have been obtained from measurements of scale atom models. In the following (a. o. Table 6) we will use the data obtained from surface chemical techniques.

peptide groups in the protein, PANKHURST suggested that association is caused by ion/dipole interaction (between the ionised group of the detergent and the peptide groups of the protein. The fact that some other detergents (e. g. alkylsulfonates) show the same maximal binding value speaks in favour of this hypothesis. In other cases, however, the maximal binding has been found to be definitively lower. There are even cases where only equivalent binding could be found by PANKHURST. PANKHURST noticed that a correlation exists between the spreading area of the detergents and the deviation from the expected maximal value. The larger the surface covered in a monomolecular layer, the smaller the maximal binding to gelatin.

This led to the additional hypothesis that the detergents having a large spreading area would be prevented from approaching the peptide groups by the side chains of the protein. Compare Table 5, in which the possibility of approach of the polar head of the detergent to the peptide group of the protein has been given in per cent penetrability. In this table we also find the spreading area per molecule detergent as given by PANKHURST (it should be noted that according to SCHULMAN (1949) the area for cetyltrimethylammoniumbromide is too high and should be replaced by the value given in brackets).

Though in general the correlation noticed by PANKHURST remains after this correction, we have found some facts which cannot be easily reconciled with PANKHURST's hypothesis. On the one hand the maximal binding of oleate to gelatin found experimentally would point to the impossible value of 150% for the penetrability (see § e), on the other hand it proved possible to realise the coacervation between gelatin and cetyltrimethylammoniumbromide, which, according to PANKHURST, would not be possible in consequence of the too large polar head (penetrability $= 0$, see Table 5).

c) The rôle of the salt necessary for coacervation

PANKHURST showed that the binding detergent/gelatin is increased by salt and that a maximum is reached above a certain concentration of salt. The rôle of the salt has not, however, been elucidated.

Directly after we were aquainted with the coacervation oleate-gelatin-K-tetraborate we were intrigued by the question how coacervation could come to the fore under these circumstances. A normal complex coacervation could be excluded as—at pH $= 9.2$—both gelatin and oleate are negatively charged. Perhaps it would be a tricomplex coacervation (compare chapter 4), where the amphionic nature of the protein is of the utmost importance. We remind of the fact that tricomplex formation is possible when the colloid anion has a sufficient affinity to the positive groups of the amphion, while the added cation must have a strong affinity to the negative groups of the amphion. Moreover the affinity between cation and colloid anion should be low (see page 37).

At first sight the coacervation gelatin $+$ oleate $+$ K-salt resembles a tricomplex coacervation very much. Any combination of two components does not lead to coacervation (at least at low K-concentrations); all three components are required.

On closer consideration of the sequences of the cation affinities of gelatin on the one hand and oleate on the other, however, we may conclude that the chances for a pure tricomplex formation are rather small. In both cases we are dealing with COO$^-$-groups and in both cases the sequence of affinity is:

$$\frac{Mg}{2} > Li > Na > K \text{ (gelatin)}$$

$$\frac{Mg}{2} > Li > Na > K \text{ (oleate).}$$

Thus affinity of the cations to the colloid anion (oleate) must be comparable to that of the cations to the amphion (gelatin). This fact is already unfavourable for the idea of a pure tricomplex formation.

Should tricomplex formation play a rôle, then we would expect that the concentration of the chlorides required for coacervation would increase in the direction:

$$\frac{Mg}{2} < Li < Na < K.$$

This sequence can only partly be put to the test, as oleate flocculates with Mg and Li. The experiment (Fig. 115) is not in contradiction with the expectation (Bungenberg de Jong and Mallee 1952).

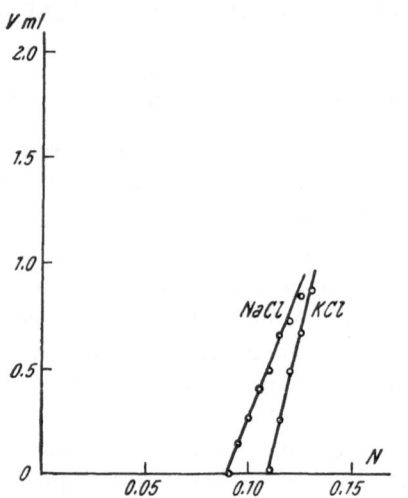

Fig. 115. Coacervation of gelatin + oleate with NaCl and KCl at pH 9.2. V = coacervate volume.

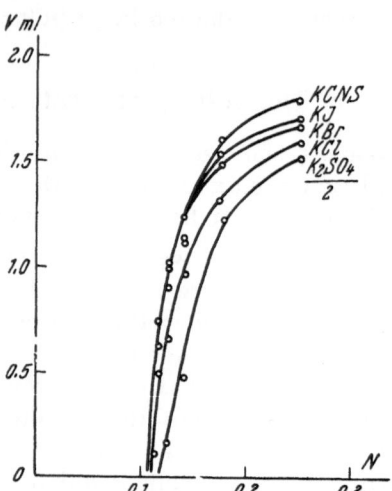

Fig. 116. Coacervation of gelatin + oleate with a number of K-salts at pH 9.2. V = coacervate volume.

As only the cation of the salt is concerned in the tricomplex interaction, we would further expect that the anion of the salt plays only a minor rôle. Here too the experiment (Fig. 116) is not in contradiction with the expectation. At higher salt concentrations we actually find differences between the anions, but coacervation starts at approximately the same concentration in all cases.

Thus we do not find definite arguments against the idea of tricomplex interaction. On the other hand, however, the experiments may not be seen as definite arguments in favour of the hypothesis, as the cation sequences for gelatin and oleate are the same.

From the foregoing it will be clear that the question—tricomplex formation or not—may be solved only when we choose a combination where the cation sequences for detergent and gelatin are opposite (this we find in the case of alkylsulfate/gelatin). In Fig. 117 the results of such an investigation have been plotted. As regards the anions we see in Fig. 118 that coacervation starts at the same concentration, in other words, the anions play a secondary rôle only. The cations, on the other hand, show an important divergence as regards their action.

The sequence of cations in Fig. 117 is (salt concentration needed for coacervation):

$$\frac{Ca}{2} < \frac{Mg}{2}; \quad K < Na < Li.$$

As the cation is bound to the COO$^-$-group of gelatin in the case of tri-complex formation, one would expect the same sequences for the coacer-

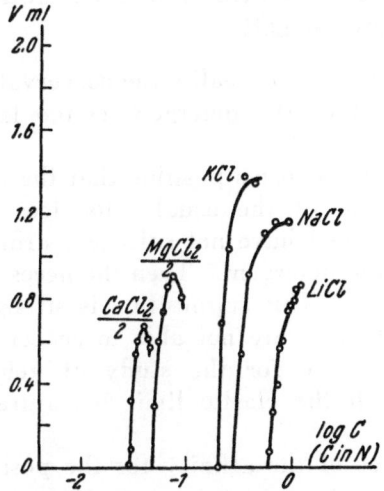

Fig. 117. Coacervation of gelatin + T-pol with CaCl$_2$, MgCl$_2$ KCl, NaCl and LiCl. V = coacervate volume in millilitres.

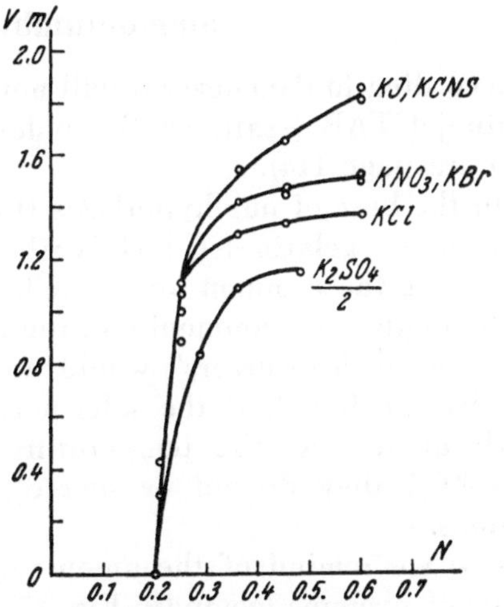

Fig. 118. Coacervation of gelatin + T-pol with a number of K-salts. V = coacervate volume in millilitres.

vation of oleate + gelatin + salt and for alkylsulfate + gelatin + salt (viz. Li < Na < K).

From the fact that in the latter combination the reverse order is found, we must conclude that tricomplex binding is not important in these coa-cervates. *The cation sequences found are the same as those for oleate and alkylsulfate respectively.* It must be concluded that the binding of the cations to the detergents is a most important factor in these interactions.

As we have found the same cation sequences for the formation of viscous-elastic systems or coa-cervates for the detergents alone, we come to the following supposition as regards the coacervation gelatin + detergent + salt.

According to this hypothesis the added salt leads to the formation of small sandwich-micelles ("pre-cursors"). Only these small sandwich-micelles may associate with the linear protein molecules. This binding gelatin/precursors contributes to the merging of the precursors in large sandwich-micelles. So we come to the picture of a large sandwich-micelle coated on both sides by a monomole-

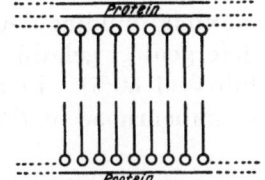

Fig. 119. Scheme for the association of long chain ions and linear protein macromo-lecules (at a sufficient salt con-centration). See text for the shortcomings in this scheme.

cular layer of linear protein. The diagram given previously (Fig. 119) is incomplete, as the oppositely charged ions of the added salt, which have caused the formation of the precursors, have not been represented. Moreover, the question of the binding between the monomolecular layers and the sandwich-micelle had not been answered. Both points will receive our attention later on.

d) The realisation of the coacervation gelatin + cetyltrimethyl-ammoniumbromide + salt

According to PANKHURST it will not be possible to realise the coacervation gelatin + CTAB + salt, as the polar head of the detergent is too large (compare page 114).

On the base of our hypothesis (Fig. 119) it seems possible that the coacervation of gelatin + CTAB with the aid of the usual salts does not succeed, as the common anions (Cl, NO_3, etc.) have not sufficient affinity for the quaternary ammonium group of the detergent. Then the necessary formation of "precursors" would not occur. This supposition is strengthened by the fact that the salts normally used are not able to coacervate CTAB alone. At the temperature necessary for the study of gelatin (35—40⁰ C.) they do not or scarcely reach the elastic limit in saturated solutions.

Now we remind of the anion sequence of the affinity for the positive groups of proteins (compare Fig. 17):

$$CNS > J > NO_3, \; Br > Cl.$$

It might be possible that the ions at the left would be able to perform the formation of percursors. It appeared that KCNS and KJ actually gave elastic-viscous systems and coacervates with CTAB alone.

Thus it was not surprising that coacervation in the system gelatin + + CTAB + salt appeared to be possible at a CNS⁻ equilibrium concentration of 10 millimoles/l. (BUNGENBERG DE JONG and WEIJZEN 1954). The maximal binding detergent/gelatin has been found to be about 6 millimoles detergent/g. gelatin (according to PANKHURST this would mean a penetrability of 60%). In our opinion the value found experimentally is in the neighbourhood of the expected maximal binding (see later § g).

e) The coacervation gelatin + oleate + K-salt

This coacervation can only be accomplished at high pH, as at lower values of pH complications arise in consequence of the formation of "acid soaps" (BUNGENBERG DE JONG, MALLEE und RECOURT 1952; BUNGENBERG DE JONG and MALLEE 1953; BUNGENBERG DE JONG, VAN SOMEREN, and KLEIN 1954). We used K-tetraborate, which salt played the double rôle of a buffer (pH = 9.2) on the one hand and of the coacervating agent on the other. The investigations have been performed with two kinds of gelatin (different isoelectric points: 5.0 and 9.2 respectively). The isoelectric points

have been determined electrophoretically (modified NORTHROP cuvette)—
Figs. 120 and 121.

We will first discuss the experiments with the gelatin having an i. e. p.
of 5.0. In Fig. 122 we have plotted the coacervate volumes obtained from a
gelatin solution and an
oleate solution in vary-
ing proportions (curves
at constant K-tetraborate
concentration). It may
be observed that coa-
cervation first appears
at a mixing ratio of ap-
proximately 67.5%. An
increase in K-tetraborate
concentration results in
a shift of the maximum
in the curve to higher
mixing ratio's. Finally
the top remains station-

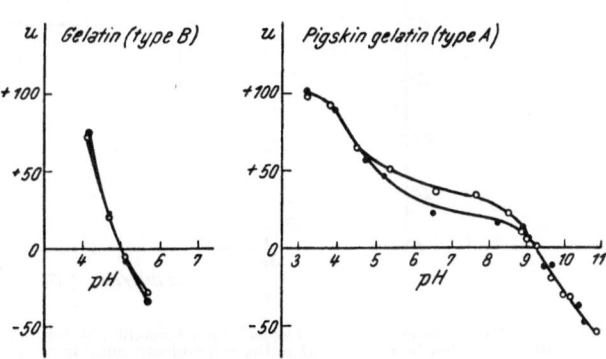

Fig. 120 and 121. Determination of the I. E. P. of the two kinds
of gelatin used in the formation of coacervates with oleate + potassium tetraborate.

ary at a certain mixing ratio and becomes independent of the salt concentration (compare Fig. 123). The horizontal branch in this diagram lies
at a mixing ratio of $80.3 \pm 0.7\%$ oleate solution. If we suppose that the
position of the top denotes the composition of the gelatin/oleate association,
one may compute from the concentration of the gelatin solution (1.7 g. dry

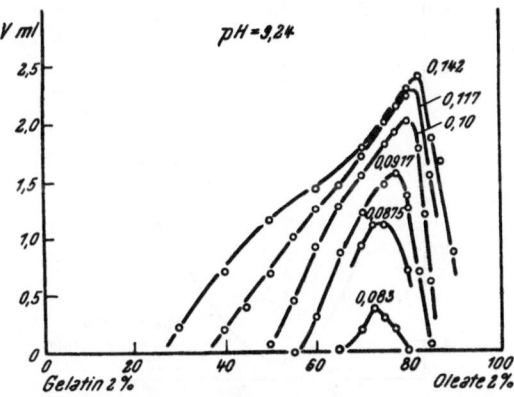

Fig. 122. Coacervate volume as a function of the mixing
ratio of a gelatin solution (1.7 g. dry gelatin/100 ml.) and
an oleate solution (65.8 millimoles/l.). I. E. P. of the
gelatin is 5.0.

gelatin/100 ml.) and the concentration of the oleate solution
(65.8 m moles/l.) a binding of
15.8 ± 0.7 millimoles oleate/g.
gelatin.

In a certain case (K-tetraborate concentration in the region of the horizontal line in
Fig. 123) the composition of the
coacervates and the equilibrium
liquids has been analysed for
a number of mixing proportions
(fatty acid determinations and
KJELDAHL-N determinations). The
results have been plotted in
Fig. 124. The curves for the
analytically determined ratio oleic acid/gelatin intersect a 15 m moles oleic
acid/gelatin. At this point—which should be regarded as the composition
of the characteristic association—the ratio oleate/gelatin is the same in the
coacervate and in the equilibrium liquid. The value found in this way
tallies satisfactorily with the curves of the coacervate volumes. There is
still another way in which we arrive at the same value, viz. the determin-

ation of the dry weight of coacervate and equilibrium liquid in a mixing series (Fig. 125).

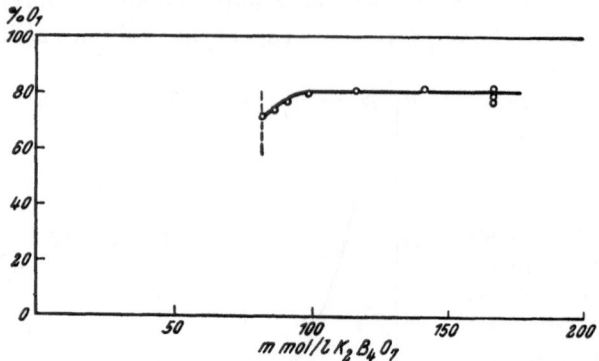

Fig. 123. Mixing ratios which correspond to the maxima in Fig. 122 and of other coacervate volume curves, not represented in Fig. 122. *O* is the same oleate solution as denoted by in Fig. 122 "Oleate 2%."

As the index of refraction of a gelatin or an oleate solution is higher than water, one might also use the maximum of refractivity of the coacervates and the minimum of refractivity of the equilibrium liquids.

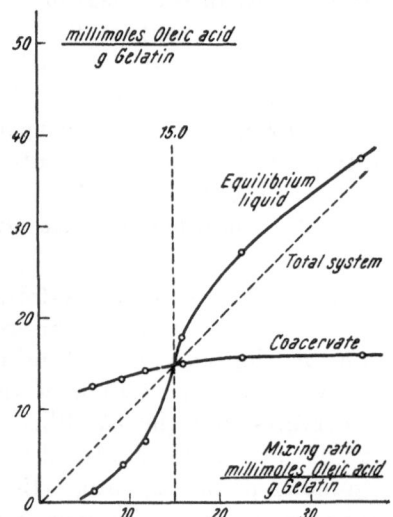

Fig. 124. Colloid composition of coacervates and equilibrium liquids in mixing series gelatin + oleate + potassium tetraborate constant (pH 9.2) as a function of the mixing ratio. Composition and mixing ratio have both been expressed as millimoles oleic acid/g. dry gelatin.

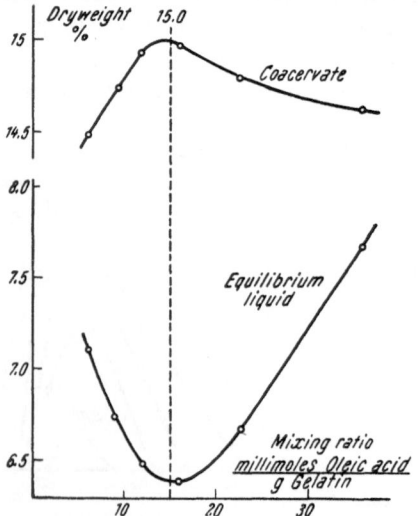

Fig. 125. Dry weights of the coacervates and equilibrium liquids in the mixing series of Fig. 124. (N. B. the high value of the dry weight of the equilibrium liquid at the minimum is due to the potassium tetraborate present.)

The other gelatin (i. e. p. = 9.2) shows a somewhat different behaviour. Here we find—at a certain K-tetraborate concentration (94 m moles/l.)—coacervate volume curves with two maxima (Fig. 126 *b*). The refractivity curve, belonging to the same salt concentration (Fig. 127), suggests that in this mixing series two coacervates of a different nature come to the fore. We come to this conclusion as we find two minima in the refractivity curve of the equilibrium liquid. The left hand coacervate is much poorer in water (higher refractivity) than the right hand one (here the presence of a separate top is barely indicated). The coacervates of both types are mutually mixable. At increase of the K-tetraborate concentration the saddle between the two tops disappears gradually (see Fig. 126 a, upper curve). A decrease of the salt concentration, however, deepens the minimum. Two separate

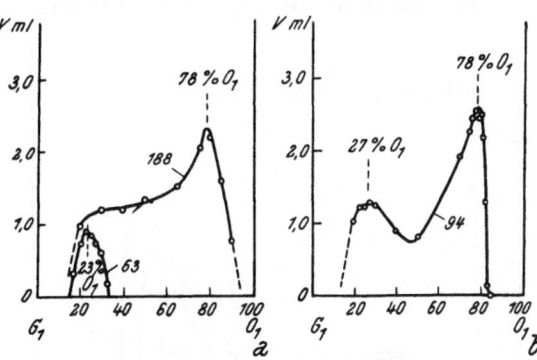

Fig. 126. Coacervate volumes separated from the mixtures of O (62.5 millimoles/l. oleate) and G (1.556 g. dry pigskin gelatin/100 ml.). The numbers indicate the concentrations of potassium tetraborate in millimoles/l. The pigskin gelatin used has an I. E. P. = 9.2.

curves will appear, of which the right hand one decreases quickly, to disappear at 83 m moles/l. The left one will only disappear at a concentration

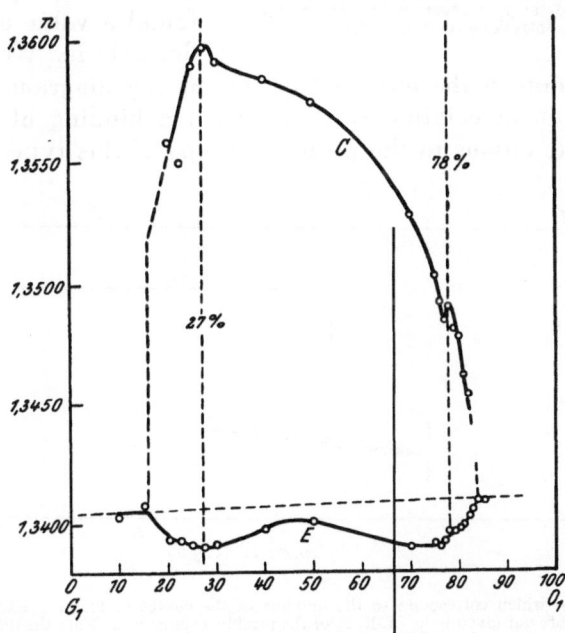

Fig. 127. Refractive index (n) of the coacervates (curve C) and of the equilibrium liquids (curve E) occurring in the mixing series which is represented in Fig. 126 b.

below 50 m moles/l. (compare Fig. 128). If we plot once more the mixing ratios of the maxima of the coacervate volumes as function of the salt concentration we obtain Fig. 129. The upper curve resembles that of Fig. 123

very much. The horizontal level is found at approximately the same value (composition of the associate 14.9 ± 0.6 m moles oleate/g. gelatin).

The difference of Fig. 129 with Fig. 123 is the presence of a second maximum (lower curve in Fig. 129). To compute the composition of this second associate it seems the best way to start with the left endpoint of this curve (Fig. 128) as this point is least influenced by the other maximum. In this way we obtain the value:

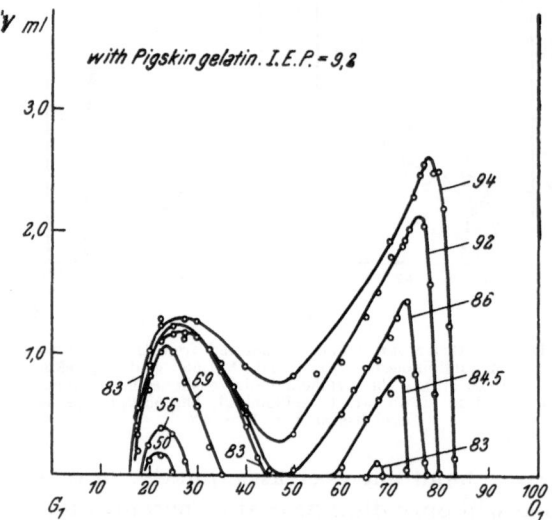

Fig. 128. Volume of the coacervate from mixtures of G, and O (see legend Fig. 126) at a number of potassium tetraborate concentrations (millimoles/l.).

1.17 ± 0.07 millimoles oleate/g. gelatin.

In our laboratory Loeven determined the total amount of basic groups of this type of gelatin (i. e. p. $= 9.2$) and found a value of 1.14 m. eq./g. dry gelatin. We may conclude that the maximum in the left part of the mixing diagram (Figs. 128 and 129) points to an association resulting from a binding of an equivalent amount of oleate anions to the positive groups of this type of gelatin.

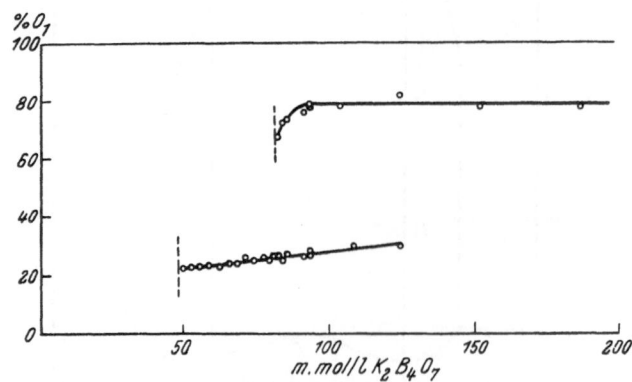

Fig. 129. Mixing ratios which correspond to the maxima of the curves in Fig. 128 and in other coacervate volume curves, which are not given here. I. E. P. of the pigskin gelatin 9.2. Note the difference with Fig. 123 in which the gelatin has an I. E. P. = 5.0.

As the other type of gelatin (i. e. p. $= 5.0$) contains practically the same amount of basic groups (1.2 m. eq./g. gelatin) one would at first sight expect that here too a second maximum in the coacervate volume curve would appear. On second thoughts, however, it is easily realised that we will not find such a maximum. All experiments have been performed at pH = 9.2. In one case (gelatin with i. e. p. = 9.2) the circumstances are ideal for tricomplex coacervation. The positive and negative groups of gelatin are balanced; every positive group may bind a soap anion and every negative group a cation. In the other case (gelatin with i. e. p. = 5) the pH of the medium is very unfavourable for tricomplex coacervation, as here the gelatin will bear a high negative charge. It is certainly possible that the binding between the soap anions and the positive groups of the gelatin does occur, but this binding has no easily visible consequences (no coacervate is formed).

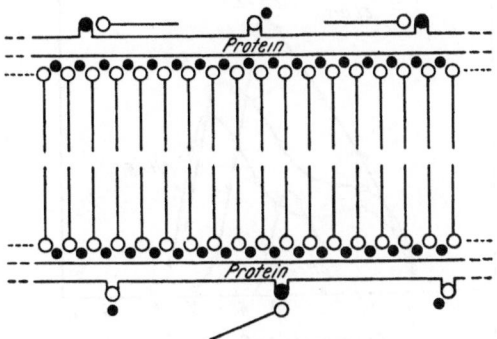

Fig. 130. Scheme of the oleate-rich gelatin-oleate association. Black dots are positively charged groups or ions, the white circles are negatively charged groups or ions.

In our opinion the equivalent binding may also remain at the right hand side of the mixing diagrams. Thus we arrive at the following scheme of the association in the oleate-rich coacervate (Fig. 130). All considerations discussed up till now have been collected in this scheme:

1. The sandwich-micelle is the centre of the association; the polar heads are connected by cations.

2. The sandwich-micelle is coated on both sides by a monomolecular protein layer.

3. In a sense the idea of an ion/dipole binding between protein and detergent (PANKHURST) has been maintained in this scheme. Then the protein monolayer must lie on the surface of the sandwich-micelle with the side chains parallel to this surface. In this way an interaction may occur between the peptide groups and the charge mosaic of the sandwich micelle.

4. The acid and basic groups at the end of the side chains of the protein will try to rise to the water and the positive groups will bind an equivalent amount of oleate ions.

f) Particulars of the coacervation gelatin + CTAB + KCNS

In the same way as has been described for the gelatin/oleate interaction, the coacervate volume as a function of the mixing proportion of CTAB and gelatin has been measured (BUNGENBERG DE JONG and WEIJZEN 1954). The pH has been varied, using acetate or phosphate buffers (see Fig. 131). The concentration of free CNS^- has been kept constant at 10 m moles/l. The computed mixing ratio at the equivalent point between the CTA-cations and the COO^--groups of the gelatin has been given by dotted lines in Fig. 131. We always find at this point an indication of a maximum, be it a S-shaped bend in the curve or just a maximum. Here too we must suppose that tricomplex coacervation occurs. This is very well possible as the experiments have been performed with the gelatin having an i. e. p. of 5.0, while the pH's of these experiments are in the neighbourhood of this value.

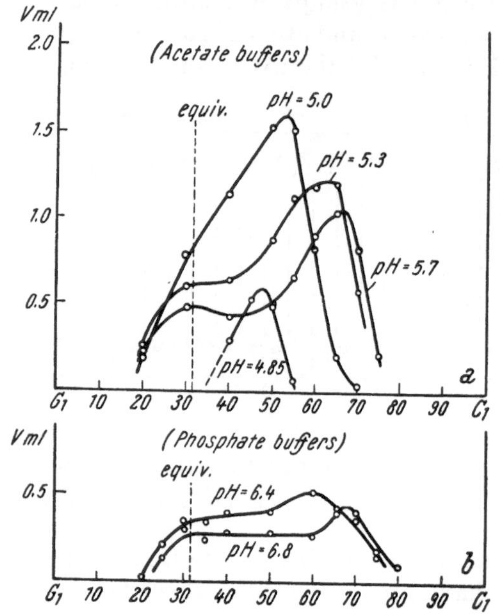

If we compute the binding between CTAB and gelatin from the position of the maxima in Fig. 131 we find the values plotted in Fig. 132. It is clear that an increase

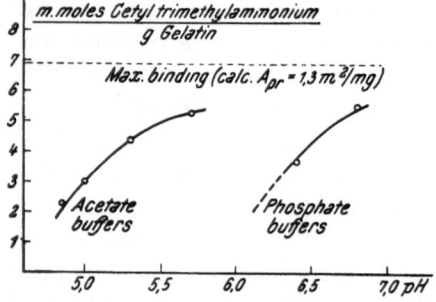

Fig. 131. Influence of the pH on the coacervation of ½% Cetavlon and ½% gelatin in the presence of 10 millimoles/l. free KCNS.

Fig. 132. Diagram representing the binding of CTA ions to the gelatin as a function of the pH, from experiments with acetate and phosphate buffers. For the calculated maximum binding see below section g.

in pH increases the binding. On the other hand specific salt influences come clearly to the fore (difference between acetate and phosphate).

To investigate the influence of added salt a series of experiments have been performed on the effect of NaCl (Fig. 133, blank, 25, 50 and 100 m moles NaCl/l.). It is seen that the top of the blank curve is displaced to the right by 25 m mol/l. NaCl. Then, at higher salt concentration the top shifts to the left, while at 100 m mol/l. NaCl coacervation no longer appears. Thus at low NaCl concentration the binding CTAB/gelatin increases, to decrease at higher concentrations. In Fig. 134 the influence of several salts has been plotted (here the ratio

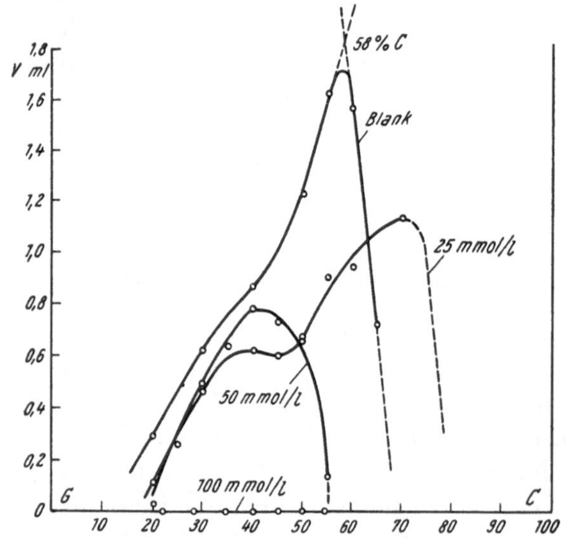

Fig. 133. Influence of NaCl on the coacervation of gelatin with CTAB in the presence of 10 millimoles/l. free KCNS. V = coacervate volume.

CTAB/gelatin has been kept constant, while 10 m mol/l. free KCNS is present).

We observe that all salts make the coacervation disappear. The salts arrange themselves according to the *double valency rule*:

$$3–1 > 2–1 > 1–1$$
$$1–2 > 1–1$$

In the immediate neighbourhood of the blank point we find a temporary *continuous valency rule*.

$$3–1\cdots2–1\cdots1–1\cdots1–2$$

relative   relative
positivation   negativation

The appearance of these two valency rules strongly indicates that in this coacervation COULOMB-forces play a rôle.

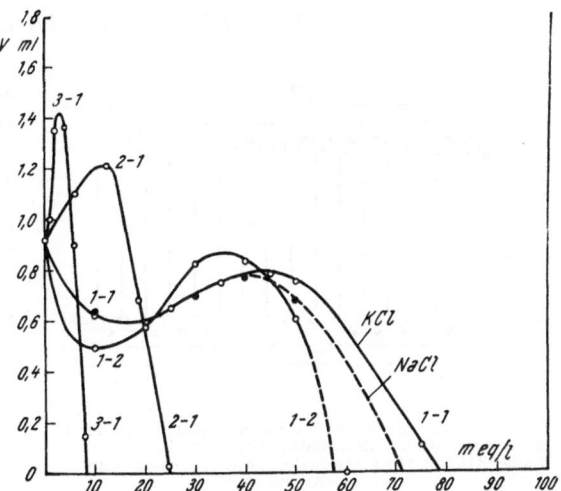

Fig. 134. Influence of salts on the coacervation at a constant mixing ratio of gelatin + CTAB in the presence of 10 m mol/l. free KCNS. 3—1 = luteocobaltic chloride, 2—1 = calcium chloride, 1—1 = KCl and NaCl and 1—2 = potassium sulfate.

From the data available it is possible to deduce the change in the ratio CTA/gelatin under the influence of added salts (Fig. 135). The influence of K_2SO_4 (1—2) is comparable to that of NaCl (see also Fig. 134). At first the binding CTA/gelatin is increased, later it decreases. The salts $CaCl_2$ (2–1) and $Co(NH_3)_6Cl_3$ (3–1) diminish the binding from the beginning.

So we must conclude that even in the CTA-rich coacervates (the highest maximum in Fig. 135 denotes a binding of 6.2 millimoles CTA/g. gelatin) COULOMB-interactions play an important rôle. We may then ask where the "point of attack" of the salts is situated.

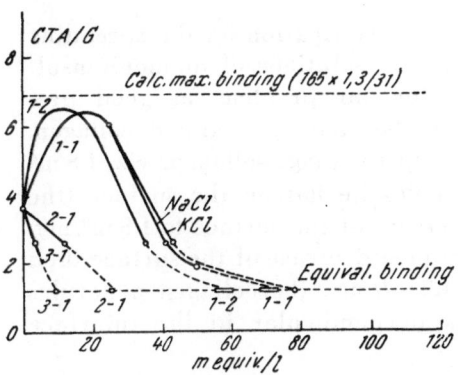

Fig. 135. Influence of salts on the CTA/gelatin ratio of the gelatin-CTA-CNS association (expressed in millimoles CTA/g. dry gelatin).

Two possibilities might be suggested: a) on the sandwich-micelles themselves, which are then regarded as complex systems, and b) on the binding of the protein monolayers to the sandwich-micelle (which implicates that this binding is caused also by COULOMB forces).

A special investigation on the influence of the salts of the types 3–1, 2–1, 1–1 and 1–2 on the coacervation and the formation of elastic-viscous systems of CTAB and KCNS (in the absence of gelatin) permits the exclusion of possibility a). Some influence has been found, but the effects are far too small to explain the influence of salts observed in the coacervation gelatin + CTAB + KCNS.

So we arrive at the conclusion that the primary point of attack of salts is the binding of gelatin to the sandwich micelle. In this case the protein monolayer must be bound by Coulomb forces. We then arrive at the scheme pictured in Fig. 136. We suppose that the negative groups of the protein play an important rôle in the binding to the sandwich micelle.

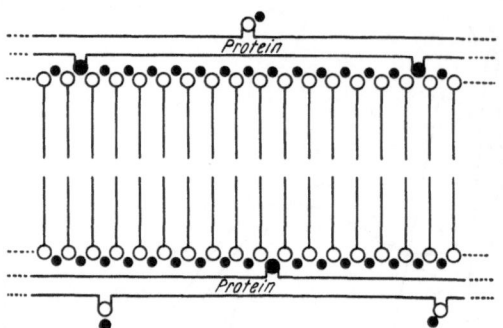

Fig. 136. Scheme for the cetyltrimethylammonium-rich-gelatin association. Black dots: negatively charged groups or anions (CNS), the white circles: positively charged groups or cations.

The study of the gelatin + CTAB + KCNS coacervate thus led to the conclusion that we may distinguish two types of detergent-rich protein/detergent associations (compare Figs. 130 and 136). In the case of oleate there are no indications of Coulomb interactions and we must suppose that the binding is caused by ion/dipole forces. In the case of CTAB, however, the Coulomb forces play an important rôle.

g) Structure and composition of the detergent-rich associations

Ellis and Pankhurst (1954) published an investigation on the spreading of collagen from a solution in formic acid on solutions of ammoniumsulfate. Their results are very important for our problem, as good preparations of gelatin may be considered to be slightly modified collagen. Two characteristic surfaces may be taken up by 1 mg. collagen, viz. 1.8 m^2 and 1.3 m^2. Originally the protein molecules lie flat on the surface (the side chains parallel to the surface). At decrease of the surface to 1.8 m^2/mg. the molecules will make contact. At continued decrease of the surface their orientation will change. Eventually, at 1.3 m^2/mg., the collagen molecules lie closely packed with the side chains perpendicular to the interface water/air.

These results, combined with our experiences that two types of gelatin/detergent associations exist, led to the idea that one type has the configuration A (in Fig. 137) and the other the structure B (Bungenberg de Jong and Weijzen 1954). Scheme A then applies to the association oleate + gelatin (no Coulomb forces, see Fig. 130) and B to the association CTAB + gelatin (Coulomb forces between protein and sandwich micelle, see Fig. 134).

If our ideas on the structure of these associations contain some truth, it must be possible to compute the maximal binding capacities from data on the spreading of the protein (in one of its two positions) and of the detergent. This maximal binding will be:

$$165.0 \times \frac{A_{\mathrm{pr}}}{A_{\mathrm{lch}}} + B \text{ millimoles detergent/g. gelatin.}$$

In this formula:

A_{pr} = the surface (m²) taken up by 1 mg. linear protein in one of its characteristic positions,

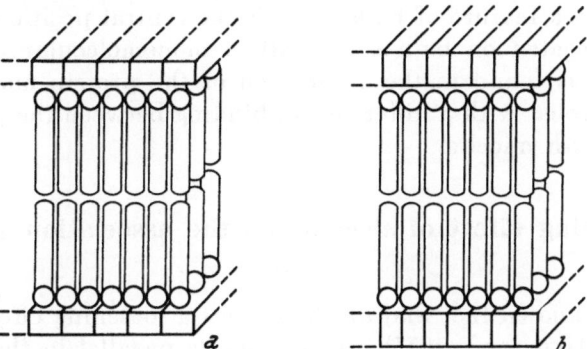

Fig. 137. Scheme for the two kinds of associations of gelatin + long chain ions.

A_{lch} = the surface (Å²) taken up by one detergent molecule,

B = number of oppositely charged groups in the protein in milliequivalents/g. protein (to this group an equivalent amount of detergent-ions will be bound; in the second case we find the value zero).

Table 6.

Associations of gelatin with	A_{lch} in Å²	Maximal binding in millimoles/g. gelatin		
		experimentally determined	computed	
			$A_{\mathrm{pr}} = 1.3$ $B = 0$	$A_{\mathrm{pr}} = 1.8$ $B = 1.2$
Oleate	20.5	15.2	10.5	15.7
Prim. alkylsulfate	20.5	10 ± 0.5	10.5	15.7
Cetylamine	20.5	9 ± 1	10.5	15.7
Cetyltrimethylammonium . . .	31	6.2 ± 0.3	6.9	10.8
7-alkylnaphthalene-3-sulfonate .	35	6.2	6.1	9.7
Sec. alkylsulfate	41	4.6	5.2	8.4

As we have no data on the spreading of gelatin we will use for A_{pr} the values found by ELLIS and PANKHURST for collagen. In doing this we introduce a systematical error. We cannot say a priori whether this error will be large or small.

From our calculations (Table 6) we get the impression that the error is relatively small. We see from a comparison of columns 3, 4 and 5 that in all cases one of the computed binding capacities lies in the neighbourhood of the values determined experimentally, while the other differs very much. Moreover, our surmise that the oleate/gelatin association belongs to type *A*, while the CTAB/gelatin associate belongs to type *B* is corroborated by the values brought together in the table.

The other four binding capacities have been found experimentally by Pankhurst. In view of the satisfactory agreement with the values in columns 3 and 4 they all belong to type *B*.

In conclusion it may be said that two types of associations presumably exist. A common feature of both types is the central position of the sandwich micelle, coated on both sides with a monomolecular protein layer. The difference is found in the orientation of the protein molecules. This is closely connected with a difference in binding between the protein layers and the sandwich micelle.

h) The loosening effect of alcohols on the association gelatin + detergent

It is a well known fact that alcohols exert a loosening effect on natural lipoproteins (Macheboeuf 1953). This finds its parallel in the suppression of the coacervation gelatin + detergent + salts by alcohols (Bungenberg de Jong and Booij-Van Staveren 1942; Bungenberg de Jong and Van Leeuwen 1952; Bungenberg de Jong and Recourt 1954). As we have already seen

Table 7.

System	Salt-sparing influences (coacervation promoted)	Salt-demanding influences (coacervation suppressed)
	number of C-atoms of the alcohols	
CTAB + KCNS	10–9–8–7–6–5	4–3–2–1
CTAB + gelatin + KCNS	10–9–8–7	6–5–4–3–2–1
T-pol (sec. alkylsulfate) + MgCl₂. . . .	9–8–7–6–5–4–3–2–1	
T-pol + gelatin + MgCl₂	9–8–7–6	5–4–3–2–1
Oleate (pH=9.2) + K-salt	8–7–6–5	4–3–2–1
Oleate + gelatin + K-salt		8–7–6–5–4–3–2–1

that coacervates of detergents alone (made with the aid of salt) are strongly influenced by alcohols it is necessary to compare the two coacervates. The following table gives a survey of our experimental results.

We see that the transition in the homologous series of alcohols always shifts when we compare detergent alone with the detergent + gelatin association. This suggests that the binding protein/detergent-micelle is hindered by the alcohols.

In an endeavour to explain the effects collected in Table 7 we came to the following hypothesis. It is suggested that all alcohols act in the same direction as regards the binding gelatin/detergent-micelle. Two types of loosening influence might be supposed (Fig. 138). The alcohol competes with the gelatin molecule for the surface of the sandwich micelle (*a*) or the alcohol is taken up in the micelle and thus creates a weak spot in the detergent/protein interaction (*b*). It seems reasonable to suggest that short alcohols will act in the first way, while long alcohols exert their influence in the second way.

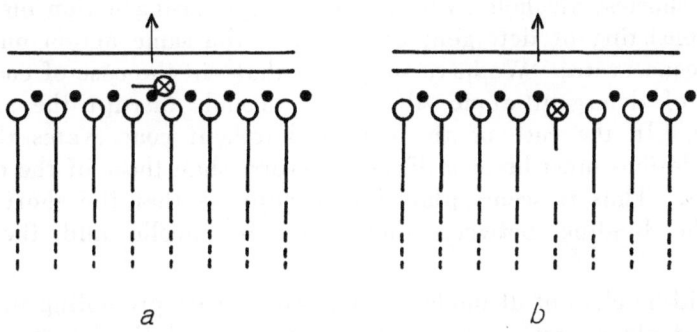

a   b

Fig. 138. Loosening influence of alcohols on the binding gelatin/detergent-micelle.

If this hypothesis is true, we would expect that the shift in transition, outspoken in the case of alcohols (Table 7), is only of small importance in the case of substances concentrating in between the methyl planes of the detergent-micelles (Fig. 81 *d*). Experiments in this direction confirm this idea. In Table 8 we see that the shift in transition from coacervate promotion to coacervate suppression is very small.

Table 8.

System	Salt-sparing influence (coacervate promoted)	Salt-demanding influence (coacervate suppr.)
	Number of C-atoms of the alkanes	
T-pol alone	5–6–7–8–9	10–11
T-pol + gelatin	5–6–7–8	9–10–11

Beside the influence of the alcohols on the binding between gelatin and the sandwich micelle we must not forget the influence of the alcohols on the micelles themselves. As we have already seen (p. 81) short alcohols may suppress an oleate coacervate. The case of this suppression is not yet known, but we have reason to suppose that these short alcohols break the intermicellar bonds, without having a strong influence on the micelles themselves. Thus the secondary association is disrupted by these short alcohols, which presumably concentrate at the surface of the micelles (Fig. 81 *b*). Longer alcohols (e. g. hexyl-alcohol) are taken up in the micelles and promote micelle growth (and thus coacervation will be possible at lower salt concentration). Especially the longest alcohols show this coacervate promoting action at low concentrations already. The large difference between the shorter and longer alcohols may also be demonstrated by the fact that only the longer ones (from amylalcohol on) give rise to P-coacervates).

When we now turn again to the gelatin/detergent coacervates, it will be clear that the influence of alcohols depends on the combination of the factors discussed. A division into three groups will facilitate the survey:

a) The shortest alcohols—which have a suppressing action on the coacervates consisting of detergent alone—have the same action on gelatin/detergent coacervates. We have suggested that—in the case of coacervates consisting of detergent alone—the short alcohols disrupt the secondary association. In the case of the gelatin/detergent coacervates the intermicellar relations must be of a different nature than those of the detergent coacervates. Thus it seems plausible to suppose that the short alcohols disrupt the binding between the detergent micelle and the protein layers.

b) A middle class of alcohols—though coacervate promoting in the case of detergent alone—get a suppressing action on gelatin/detergent coacervates. The influence on the detergent-micelles demands relatively high concentrations of alcohol. Before these concentrations have been reached the opposite effect—loosening of the gelatin—will take place. One may demonstrate that the detergent-micelles are still intact, as at certain (higher) concentrations of the alcohols P-coacervates come to the fore.

c) The longest alcohols—having a coacervate promoting influence on detergent alone remain coacervate promoting, notwithstanding the fact that the gelatin will gradually be loosened from the detergent-micelles. This is a result of the fact that the coacervate promoting influence on the detergent alone will be found at low alcohol concentrations already. It goes without saying that there too P-coacervates will appear at higher concentrations.

At the end of this chapter we must point out that the results described seem to be of great importance for biology. The structure of that class of natural lipoproteins which consist of linear proteins and phosphatides is presumably the same as that of the simple detergent/gelatin micelles. The association—spread in a plane—may serve as a membrane at external or internal surfaces of the protoplasm, or form three-dimensional structures by piling up. The phosphatide molecule is, as it were, predestined to form sandwich micelles, as it has a positive as well as a negative group strongly fixed by covalent bonds. This means that in contradistinction to the simple detergents, no salt is required in the medium for the formation of micelles. The presence of phosphatidic acid (or "unbalanced" phosphatides) would be the cause of weak spots in the micelle, which may be strengthened by cations. Here the possibility for biological regulation is clearly indicated.

References

BUNGENBERG DE JONG, H. G., and C. H. BOOIJ-VAN STAVEREN, 1942: Lipophile protein-oleate coacervates and the effect on them of alcohols. Proc. Ned. Akad. Wetensch. Amst. **45**, 601—606.

— and A. M. VAN LEEUWEN, 1952: Contributions to the problem of the association between proteins and lipids II. Proc. Kon. Ned. Akad. Wetensch. Amst. **B 55**, 317—346.

— — and C. MALLEE, 1952: Contributions to the problem of the association between proteins and lipids IV. Proc. Kon. Ned. Akad. Wetensch. Amst. **B 55**, 360—372.

— — 1953: Contributions to the problem of the association between proteins and lipids V. Proc. Kon. Ned. Akad. Wetensch. Amst. **B 56**, 203—217.

— — and A. RECOURT, 1952: Contributions to the problem of the association between proteins and lipids III. Proc. Kon. Ned. Akad. Wetensch. Amst. **B 55**, 347—359.

— and A. RECOURT, 1954: Contributions to the problem of the association between proteins and lipids VIII. Proc. Kon. Ned. Akad. Wetensch. Amst. **B 57**, 204—214.

— C. R. VAN SOMEREN, and F. KLEIN, 1954: Contributions to the problem of the association between proteins and lipids VI. Proc. Kon. Ned. Akad. Wetensch. Amst. **B 57**, 1—19.

— and W. W. H. WEIJZEN, 1954: Contributions to the problem of the association between proteins and lipids VII. Proc. Kon. Ned. Akad. Wetensch. Amst. **B 57**, 192—203.

— and W. W. H. WEIJZEN, 1954: Contributions to the problem of the association between proteins and lipids IX. Proc. Kon. Ned. Akad. Wetensch. Amst. **B 57**, 285—310.

ELLIS, S. C., and K. G. A. PANKHURST, 1954: Monolayers of collagen. Trans. Faraday Soc. **50**, 82—89.

MACHEBOEUF, M., 1953: Lipoproteins of horse plasma and serum. In Tullis (Ed.), Blood cells and plasma proteins. New York, p. 358—377.

PANKHURST, K. G. A., 1949: Formation of complexes between gelatin and sodium alkyl sulphates. In Surface Chemistry, London 1949, p. 109—118.

— 1949: The adsorption of paraffin chain salts to proteins V. Disc. Faraday Soc. **6**, 52—58.

SCHULMAN, J. H., 1949: Disc. Faraday Soc. **6**, 58—59.

9. Biological membranes

a) Introduction

It is a well known fact that the contents of living cells differ very much from the medium. Muscle cells, e. g. contain ten times as much K^+-ions than blood, but their Na^+-concentrations are ten times smaller. These large differences suggest two things. In the first place the living cell must have the disposal over a mechanism which allows the—selective—uptake (or secretion) of ions. The energy for this process must be derived from metabolism. Secondly there must be a barrier against the free diffusion of ions. Otherwise the ions taken up would be lost very soon. Many experiments suggest that the outer membrane of cells has the properties sought for. The tonoplast of the plant cell and the membranes between nucleus and cytoplasm or between mitochondria and cytoplasm may have comparable properties.

The problem of the structure of the protoplasmic membrane has inter-

9*

ested many investigators. Is this membrane [28] an independent part of the cytoplasm? It might also be simply that part of the cytoplasm which happens to be at the surface of the cell at a certain moment. An indirect way to gain some knowledge about the protoplasmic membrane consists in measuring the velocity of permeation into living cells of various substances. These experiments gave rise to two conflicting ideas:

1. the permeating substance dissolves in the membrane,

2. the permeating substance finds its way through pores in the membrane.

The study of permeability led to the conclusion that the balance between the hydrophobic and hydrophilic groups in a molecule has a strong influence on its velocity of penetration. Generally speaking, the more hydrophobic the molecule, the quicker the penetration into living cells. Thus we see that the velocity of penetration increases considerably in the direction glycerol, dihydroxypropane, propylalcohol. These kinds of experiments led OVERTON (1895) to his *lipoid theory*. According to this hypothesis the outer membrane of living cells consists of a layer of lipids. Each substance with a large tendency to dissolve in lipids will show a quick penetration into cells.

On the other hand it was asked by RUHLAND (1925) why substances like glycerol—practically insoluble in lipids—penetrate still rather quickly into living cells. His conclusion was that many organic molecules (especially hydrophilic ones) may penetrate through pores in the protoplasmic membrane (*filter theory*).

These two extreme rules, however, are not obeyed by living organisms. In general one might say that small hydrophilic molecules penetrate quicker than would be expected from the lipoid-theory. On the other hand, the velocity of permeation of large hydrophobic molecules does not fit into the filter-theory. These discrepancies led COLLANDER (1933) to frame his *lipoid-filter-theory*. This theory states that the protoplasmic membrane combines both principles mentioned, though it does not give an indication of the structure of the membrane.

Everyone who tries to give a general theory on the structure of the protoplasmic membrane encounters as a first difficulty the necessary combination of the two at first sight conflicting principles. Such a general theory would have to explain a lot of other difficulties. Every kind of filter theory has much trouble with the fact that ions do not (or slowly) penetrate, while e. g. glycerol does enter at an appreciable rate. Moreover, the filter theory tacitly assumes that the membrane has a more or less rigid nature. But it should be noted that the membrane must be a dynamic structure. The classical experiments by NAEGELI (1881) on root cells of *Hydrocharis* show that injured protoplasm forms a new membrane rapidly. Thus it is quite a question that this membrane has a rigid frame. The recovery of the membrane depends on the surrounding medium—especially

---

[28] Which must not be confused with the cell wall of plant cells, consisting of cellulose.

as regards the ions. Calcium ions promote the recovery, while sodium ions have an opposite action. From these experiments it must also be concluded that a strong relation exists between the cytoplasm and its membrane. All organic components of the membrane must be supplied immediately by the cytoplasm. This suggests that these materials are present in ample quantities in the cytoplasm.

The phenomenon of phagocytosis also suggests that the protoplasmic membrane must be a supple structure. Otherwise the permeability of a leucocyte would alter drastically when a bacterium is taken up. Here we approach a problem which lies on the borderland of phagocytosis: do (some) proteins enter into (some) cells? It is difficult to explain the action of hormones of protein nature without this supposition. Thus we arrive at a serious dilemma: we are looking for a structure which in some cases will let proteins pass, but which does not permit the penetration of small inorganic ions.

Then, a general theory of permeability will have to give an explanation of the fact that the permeation of relatively large hydrophilic molecules shows a high temperature coefficient. As DAVSON and DANIELLI (1943) have already pointed out, these high values of Q_{10} do not imply that chemical reactions are responsible for this transport.

Finally, the permeability differs from organism to organism and from cell to cell. A theory should not only try to give an interpretation of the experimental facts common to all cells, but also give the reasons why such a large variation in details is possible.

b) Models of the protoplasmic membrane

From the beginning the students of biological permeability have tried to find physico-chemical explanations of the variations in penetration velocity of molecules into living cells. Thus, numerous models of the protoplasmic membrane have been constructed and studied. Mostly, each of these models only elucidated one side of the complex problem and practically completely ignored the other sides. A very short survey of the most important models will enable us to see the weak and the strong points of each model.

1. The system oil/water

Though not being a membrane, the study of this system has had an enormous influence on the theoretical aspects of permeability. The fact that a rather strong correlation could be found between the distribution coefficient oil/water and the permeation velocity of many molecules has led to the lipoid theory. Of course, olive oil—which has been used in many cases—can be compared only superficially with the lipid components of the protoplasmic membrane. Here the experiments of NIRENSTEIN should be mentioned. He tried to find a more close correlation between the distribution coefficients and the permeability data—especially of dyes—by adding substances to the olive oil. It is important to note that he succeeded in this by adding oleic acid and diamylamine to olive oil. He concluded that positive and negative groups should be present in the cell surface. This constitutes an indirect argument for the opinion that phosphatides—having both negative and positive groups—play an important rôle in the protoplasmic membrane.

2. Porous "membranes"

The need was felt to have a model which is really a membrane. A first, though very crude, approximation of natural membranes is found in collodion membranes. But, when studying the penetration of organic molecules through these membranes, Collander (1926) did not find a correlation between the velocity of permeation and the oil/water partition coefficient. The molecular volume showed to be the determining factor for penetration. Choosing for his experiments more "natural membranes" (gelatin), Collander neither did find a correlation between solubility into lipids and velocity of penetration. These rather disappointing experiments gave an important argument against the view that the outer membrane of protoplasm would consist solely of proteins.

Another aspect of permeability—the influence of the charge on permeation—has been demonstrated by Michaelis. A negative membrane may inhibit the penetration of anions. This restricted mobility of the ions may result in a membrane potential. Thus these rigid porous membranes—which did not contribute much for our understanding of biological permeability—have been selected as a much studied model for bioelectric potentials.

3. Liquid "membranes"

The artificial porous membranes have a strong rigidity. As models they have not been very successful, as we have seen that the protoplasmic membrane cannot be very rigid. Then it is tempting to experiment with non-rigid (liquid) membranes. Here we think of the experiments of Osterhout (1932). He separated two compartments containing water and ions by a "membrane" of guajacol and p-cresol. Bubbling CO_2 through one of the compartments may result in a transport of cations (e. g. K^+) to that side of the apparatus. This is one of the simplest models for the active transport of substances. A substance—in itself insoluble in the membrane—is combined with a "carrier"; the combination is soluble in the membrane and may be transported.

It has not been possible to make really thin membranes. Thus it is out of the question to study the passive permeation of organic molecules through these "oil membranes." Moreover it is not easy to see how a really liquid membrane could be stable at the surface of protoplasm. A large amount of work has been done on the potential differences across oil layers, but the results do not find a ready interpretation.

4. Monomolecular films

In most cases the protoplasmic membrane forms spontaneously—after injury—at the interface cytoplasm/medium (though there are indications that metabolism may play a rôle). The question arises what substances show a spontaneous concentration at interfaces. Now the interface cytoplasm/medium is essentially an interface between two liquids containing mostly water. On the other hand experimental difficulties limit the study of interfaces in the laboratory practically completely to the interface water/air. At this interface all kinds of molecules provided with a hydrophilic group and a large hydrophobic part may form monomolecular layers. The living cell contains a number of substances (phosphatides, cholesterol, etc.) which fall into this category. This suggests that the study of monomolecular layers will be of the utmost importance for the problem of the structure of the protoplasmic membrane. To mention but one example: SCHULMAN (1949) found a marked

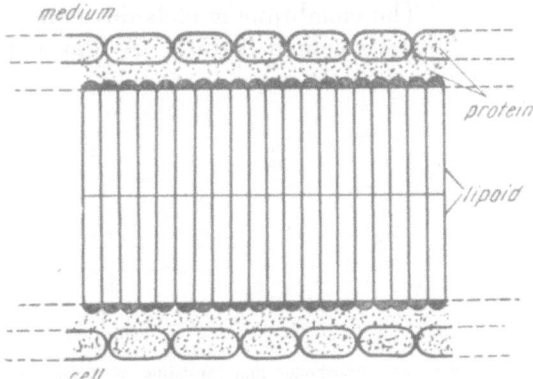

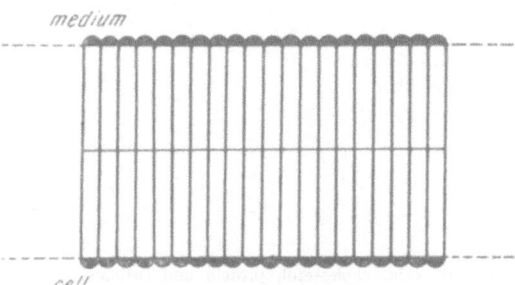

Fig. 139. Membrane model according to
GORTER and GRENDEL

Fig. 140. Membrane model after DANIELLI.

resemblance between the penetration of certain detergents in monomolecular films and their hemolytic power.

Many years ago GORTER and GRENDEL (1926) framed the hypothesis that the membrane of the erythrocyte consists of a double layer of lipids (Fig. 139). From experiments on the spreading of lipids extracted from erythrocytes they computed that the surface covered by these lipids was twice as large as the surface of the erythrocytes.

DANIELLI supposed that the lipids in the model of GORTER would have a high interfacial tension. As lipophilic colloids—e. g. soap coacervates— show a very low interfacial tension, DANIELLI's supposition is valid only in the case of uncharged lipids. Now cells have a relatively low interfacial tension medium/cell. To compensate this supposed difficulty DANIELLI suggested that the lipids are coated with a layer of protein (Fig. 140).

Two objections may be made against the monomolecular film (and the double film) as a model of the protoplasmic membrane.

a) In order to get a coherent film at the water surface a force must be applied. The question arises as to the nature of the force in the protoplasmic membrane holding the film or double film together.

b) A monomolecular film is stable at the interface between two different phases (e. g. water/air). It may be asked how it is possible that the proposed double layers are stable at the interface water/water.

5. Complex systems

The investigations of Bungenberg de Jong on complex relations (see chapter 4) elucidate the problem of the force holding together the molecules in the supposed films. We have seen (chapter 3) that the negative biocolloids may be divided into three classes: sulfate, carboxyl and phosphate colloids. Specific cation sequences come to the fore when we study the reversal of charge of these biocolloids. The influence of cations on permeability shows the characteristics of a cation sequence of the phosphate group. This provides another argument for the idea of the participation of phosphatides in the protoplasmic membrane.

The membrane models developed in Bungenberg de Jong's laboratory all have a common feature. A film of phosphatides constitutes the lipid frac-

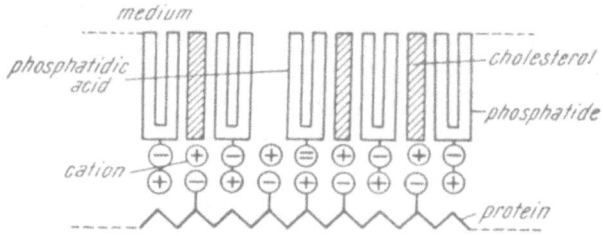

Fig. 141. Tricomplex film consisting of phosphatides, phosphatidic acid, cholesterol, protein and cations.

tion of the membrane. These phosphatides are cemented together with cholesterol. The negative charge of the membrane must be ascribed to a small amount of phosphatidic acid. Such a film would still be unstable, but it may form a tricomplex with protein and cations (see e. g. Fig. 141).

These models show two interesting features:

1. the stability of the films is due to London-van der Waals forces between the apolar parts and Coulomb forces between the ionized groups and 2. the action of ions on permeability is understood, as the model pictured is an ion-exchanger.

The main objections are:

1. the outside of the cell would be extremely hydrophobic (and show a high interfacial tension).

2. The exact equilibrium between the charged groups requires proteins of a peculiar composition.

3. The proposed exchange of cations will be very much impeded by the hydrophobic layer.

6. Lipophilic coacervates

The soap coacervates may be looked upon as models of phosphatides. As we have seen these lipophilic coacervates are very useful for the study of the hydrophilic/hydrophobic balance of organic molecules. Long after the first experiments with this model—which showed a strong correlation between the influence of organic non-electrolytes on these coacervates and on the living cell—it became clear that large flat micelles are present in the soap coacervates.

This fact is very important from a theoretical point of view, because it shows that the supposed double layer of GORTER (Fig. 139) is possible in water, provided that the ionized groups are more or less discharged. The objection against this simple model lies in the fact that in the case of soap solutions strong salt concentrations are needed to give double layers.

In the protoplasmic membrane, however, phosphatides will play an important part. In these molecules the positive groups practically compensate the negative ones. Thus micelle formation should be possible at

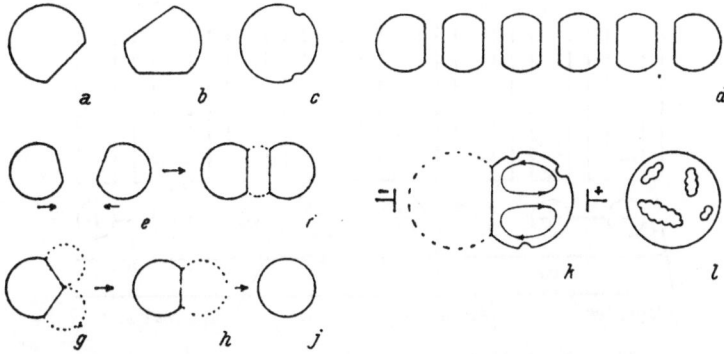

Fig. 142. *a, b, c* Drops with sideways attached vesicles, the protruding vesicle not being visible. *h-i* A mechanical disturbance may rupture the vesicle wall, whereafter the coacervate drop immediately rounds off. *e-f* Formation of a mechanical unit, consisting of two coacervate drops and a common vesicle. *d* A larger unit, consisting of several coacervate drops and vesicles. *l* Chain of vacuoles separated by invisible septa. *g-h* Sideways fusion of two vesicles. *k* Behaviour of a negatively charged coacervate drop in a direct current field. The moving boundary transports small vesicles to the cathode side of the drop, where they fuse to one vesicle which gradually increases in size. With positive coacervate drops the movement of the boundary is just the other way round and the large vesicle develops at the anode side.

very low salt concentration and the difficulties mentioned in section 4) of this chapter disappear. Theoretically speaking a double layer of phosphatides may exist at the surface of cytoplasm. When the amount of phosphatidic acid is high one would expect that cations are required for the stability of the membrane. By virtue of their strong affinity for the phosphate groups calcium ions have a strong micelle promoting action. Here we encounter possibilities of ion antagonism, as a large quantity of another salt (e. g. NaCl) may counteract the action of $CaCl_2$ (BUNGENBERG DE JONG, WAKKIE and BOOIJ 1936).

Long ago it had already been found by BUNGENBERG DE JONG and BONNER (1935) that phosphatide films may separate two media consisting mainly of water. The surface of phosphatide coacervates shows some peculiarities, which we do not find in normal soap-coacervates (see for the distinction between P- and O-coacervates p. 50). In Fig. 142 some pictures of P-coacervates are found. The remarkable flat sides of the coacervates can only be understood by the assumption that the coacervate drops have sideways attached vesicles. These vesicles contain equilibrium liquid and they are surrounded by very thin (microscopically invisible) phosphatide layers.

In the original publication it was assumed that the phosphatide molecules are oriented with their polar groups towards the coacervate. At present, however, we must leave the question of the orientation of the phosphatide molecules undecided.

As regards the problem of permeability it is very interesting to note that these experiments show that phosphatide may form membranes spontaneously under certain circumstances.

7. Synthetic "lipoproteins"

The importance of Bungenberg de Jong's work on synthetic lipoproteins lies in the fact that more became known about the possibility of interaction

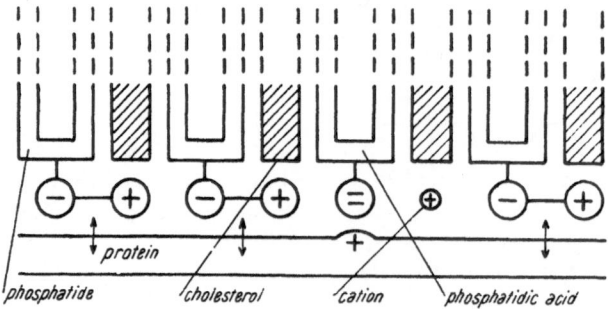

Fig. 143. Double layer of phosphatides coated with protein.

between lipids and proteins. He showed that large flat micelles of detergents coated with protein may appear even at relatively low salt concentrations (chapter 8). As regards the problem of the force drawing the protein molecules and the detergent micelles together two possibilities have been given. It may be possible that we are dealing with ion/dipole interactions between the detergent ion and the peptide link of the protein (Pankhurst). Another possibility is that the charge mosaic at the surface of the detergent micelle binds the ionized groups of the protein by complex forces. In any case it seems that the lipid micelle must be regarded as a unity (consisting of lipid molecules + inorganic ions). The great difference between the models pictured in Fig. 141 and 137 is—besides the fact that the former is a monolayer and the latter a double layer—that the first supposes an intricate (and improbable) pattern of charges connecting the phosphatide and protein layers, which is not necessary in the second one.

The binding between the two units—mixed double layer of lipids and proteins—might be illustrated in a diagram (Fig. 143), which is, of course, only one example of a great many possibilities.

It is interesting to see how two different approaches have ultimately led to the same models (compare Figs. 140 and 143). Many other researches might be cited where the idea of a double layer of lipids has come to the fore. For our problem it is very important that in cytoplasm too these layers seem to occur (see e. g. Gay and Anderson 1954). It shows that the double layers are available when they would be necessary for the repair of the protoplasmic membrane. It is even probable that a continuous exchange of material between the protoplasmic membrane and the cytoplasm takes place.

c) Theoretical aspects of permeability

A general theory of cell permeability should combine the filter principle and the solubility principle. Various theories have tried to give this combination. The *emulsion theory* need not be discussed. It is very difficult to see how an emulsion could be stable at the surface of the cell. Especially if water is the continuous phase the "membrane" will disappear immediately. The only interesting point in this theory is the influence of ions on the inversion of the phases.

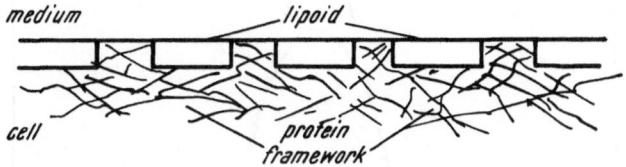

Fig. 144. The protoplasmatic membrane according to the mosaic theory.

The *mosaic theory* is far more important and has many adherents. Here the two principles are located in different parts of the membrane (Fig. 144). The difficulties arise when we try to draw this model on a molecular scale.

The only possibility which is not seriously in conflict with physicochemical laws seems to be that the lipids are embedded in a rigid framework of proteins. Then we come into trouble as many biological experiments suggest a strong flexibility of the membrane. Another difficulty is that water molecules (and ions) would pass the protein structure rapidly; in other words the cell would be like a leaky ship. Several experiments have shown that the electric resistance of cells is high and that this resistance is located at the cell's surface. A membrane with a mosaic structure could not have a high resistance because of the short-circuiting through the protein holes. Thus we are forced to suppose a continuous lipid layer at the surface of the cell.

Some years ago one of us (Booij 1949) proposed to give the name *complex theory* to the closely connected hypotheses and models developed in the laboratory of BUNGENBERG DE JONG. The link connecting these models is the suggestion that complex relations (COULOMB-forces) and LONDON-VAN DER WAALS forces tie the membrane constituents together. Thus in the latest model originated with the BUNGENBERG DE JONG school the earlier models of GORTER and of DANIELLI are now on more solid ground.

At first sight the complex theory is only one of the many lipoid theories. Booij (1954) has tried to show that the complex theory comprises the filter principle too. Drawn on paper the models (Figs. 57, 141, 143) look very static. These diagrams, however, try to give a picture of molecular dimensions, where the molecules will show an agitation depending on temperature.

Notwithstanding the parallel arrangement, the lipids in the membrane must not be seen as three-dimensional crystals; they will be in liquid crystalline state. This means that smaller and larger holes will be distributed statistically over the cell's surface. This situation will change continuously.

This point of view enables us to give an—admittedly crude—explanation of the high temperature coefficient of the permeation of large hydrophilic substances. Smaller and larger holes will be present in between the carbon chains of the lipids. These holes will be distributed according to some

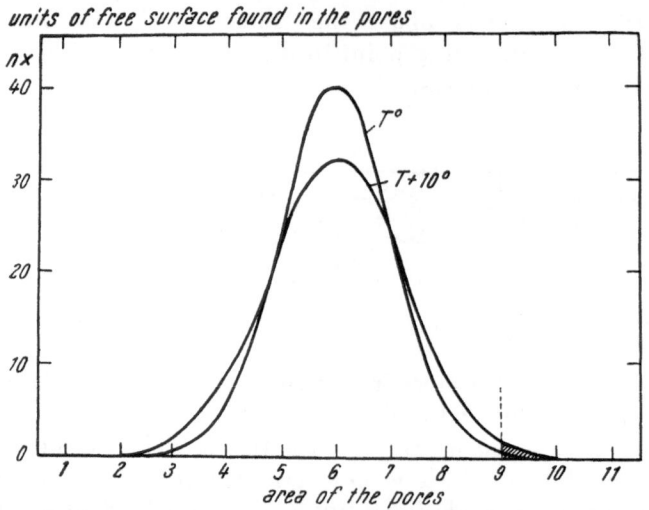

Fig. 145. Suppose we divide the surface of the protoplasmic membrane in between the carbon chains (the "free surface" into small units and determine the number of units per pore. These surface units will be distributed according to some statistical curve. A rise in temperature will cause a change in the curve which will especially influence the number of surface units found in the larger pores (shaded areas).

statistical curve (Fig. 145). Most holes will be too small to let hydrophilic molecules (even the small water molecules) pass. A rise in temperature will cause a stronger thermal agitation and the distribution changes a little. This change is very important in the region of the large holes, in other words here the temperature coefficient is large.

The complex theory in its general form contains many elements of earlier theories. It tries to reconcile these often conflicting theories. The following facts may be understood from our point of view:

a) The protoplasmic membrane behaves like a continuous layer of lipids. Lipophilic molecules will pass rapidly and a high electrical resistance will be found at the cell's surface.

b) The outside of the cell lets hydrophilic molecules pass, though slowly. Temperature influences the molecular agitation in this "statistical" filter, which results in a high temperature coefficient of the permeation of large hydrophilic molecules.

c) Changes in the medium (pH, electrolytes, etc.) have an influence on permeability—especially of hydrophilic molecules. Three effects may be mentioned:

1. The packing of the carbon chain in the lipid films may be altered.
2. the amount of water in the "electric" layer between the phosphatides and protein may be changed and 3. the binding between protein and lipid film may be influenced.

d) As in this notion the protoplasmic membrane is a flexible double film phagocytosis (and the penetration of proteins) must be seen as a spreading of the film over the foreign particle, while the integrity of the membrane rests intact. In infectious diseases phagocytosis of the pathological microbes is substantially increased by specific antibodies neutralizing the charges of the microbes.

e) The formation of a new membrane after injury to the old one means that the membrane constituents are present in the cytoplasm (perhaps as double films already). In this case the membrane is part of the cytoplasm.

f) The membrane separates two phases of completely different ionic composition. This means that the membrane too will have two different sides. This may have a strong influence on the membrane potential. Even in the case of a double layer of phosphatides this difference must come to the fore when we find at one side mainly Na-ions and at the other mainly K-ions. These ions have a different affinitiy for the phosphate groups, which must result in a membrane potential. Perhaps even the problem of the one-sided permeation may be seen from this point of view.

From this enumeration it may be concluded that the complex theory is able to give a provisional explanation of some of the phenomena which characterize biological phenomena. Such a theory, however, will not be complete without an attempt to explain the enormous biological variation which we find in the permeability of different cells and tissues. The general plan of the complex theory (the membrane consists of a double layer of lipids coated with protein and bound together by complex forces, LONDON-VAN DER WAALS forces and ion/dipole interactions [29] may be worked out in several ways. A number of factors will influence permeability:

1. As regards the apolar part of the membrane it will be clear that permeability will depend on:

a) The number of double bonds in the carbon chains of the lipids. A great number of double bonds will presumably cause a loosening of the membrane structure.

b) The proportion of phosphatides and cholesterol. It seems that more cholesterol leads to a more rigid—less permeable—structure.

c) The chain length of the fatty acid residues in the lipids. A homogeneous layer containing long fatty acid residues will be less permeable than a layer where the lengths of these residues differ very much.

2. The polar part of the membrane gives even more possibilities for variation.

---

[29] Sometimes this model is wrongly cited as a coacervate. It should be called a complex double film (for a definition of a coacervate see page 8). We study the properties of related coacervates in order to get some insight into the forces holding the double film together.

a) The nature of the phospholipids will play a very important rôle. We may point e. g. to the fact that lecithin, cephalin and phosphatidic acid differ widely as regards the influence of pH on their charge.

Lecithin will be neutral throughout the whole biological range of pH, cephalin will become negative at high pH-values, while phosphatidic acid is a monovalent anion at low pH. The large variations in the influence of pH on permeability might find their explanation in the fact that the distribution of these phospholipids among the various cells is entirely different. As many other lipids with electrolyte character exist (sphingomyelins, sulfolipids, etc.) we may expect a large variation in respect to permeability, influence of pH on permeability, influence of ions, ion antagonism, etc.

b) The protein too will vary from cell to cell. Each of the factors already mentioned will have its influence on the charge mosaic of the lipid double layer and consequently on the binding of the protein on this double layer. The distribution of the charged groups on the protein molecule will, of course, strongly influence the binding to the lipid micelle. As we have seen that there are two fundamentally different ways in which the protein layer is bound to the phosphatide micelle, we may expect that the pH will have a strong influence on the membrane of some cells, but a negligible influence on that of others. Ions, present in the medium may be bound to the protein, some even more or less specifically.

From this survey we may safely conclude that the complex theory gives ample opportunity for large deviations in cellular permeability. The number of possibilities is so large, that it will even be very difficult to prove (or disprove) the complex theory. The direct proof (analysis of the membrane constituents) is not practicable as it is not possible to isolate the protoplasmic membrane. So we will have to gather indirect arguments for or against the theory.

d) Some experiments on the problem of the protoplasmic membrane

As permeability is treated elsewhere in this handbook we will not try to give a complete survey of the experiments performed in this field. We will mention some experiments which either are difficult to explain with the aid of older theories, or have been planned in connection with colloid-chemical experiments.

Our first example will be an experiment by Green (1949) on the penetration of fatty acids in erythrocytes. He measured the velocity of penetration with the aid of a Hartridge-Roughton method. It is rather surprising to see that certain short acids show a quick penetration while the penetration of longer acids in much slower (Fig. 146). In contrast to this the longer fatty acids show a strong hemolysing action. Green suggests that, as regards penetration, the determining factor will be lipid solubility for the shorter acids, but molecular volume for the longer ones. This hypothesis does not tally well with colloid-chemical considerations. According to our view the long acids show a large solubility in lipids. The consequence will be that they stick in the double layer of lipids. Eventually the double layer is gravely disturbed and the erythrocyte hemolyses. The following table gives the background of the penetration phenomenon.

A substance with a high solubility in lipids will not pass the membrane. This is an important correction of the original lipoid theory. As we suppose a close connection between the protoplasmic membrane and cytoplasm such a substance may reach all lipid fractions of protoplasm (e. g. mitochondria); but it will not be found in the vacuoles.

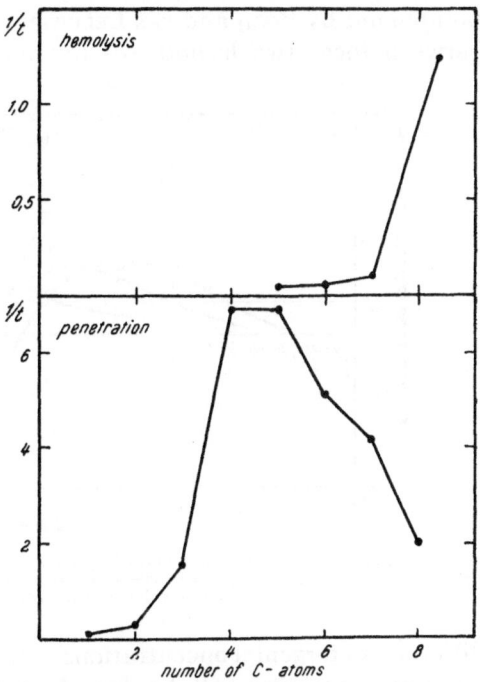

In the field of the plant growth regulators VELDSTRA and BOOIJ (1949) found a remarkable phenomenon. A solution of naphthalene acetic acid of low concentration (e. g. 10^{-6} mol/l.) shows only a very weak growth-response in the pea test. This response will be greatly strengthened by the addition of solutions of decahydronaphthalene acetic acid (2.10^{-5} mol/l.) or di-n-amylacetic acid (4.10^{-5} mol/l.). The latter substances are inactive compounds. The hypothesis given was that naphthalene acetic acid is distributed in the plant cell (it will be found in the ecto- and endoplasmic membranes and at the "centre of the primary action"—an enzyme?). If we add to an "underdosed" highly active growth substance a substance with a stronger

Fig. 146. The influence of fatty acids on hemolysis (t in minutes) and the penetration of fatty acids into erythrocytes (t in seconds) according to GREEN.

"membrane affinity" (see the experiments on the influence of these substances on the oleate coacervate) it will displace the growth substance from the lipid membranes and more will become available for the "active centre." The relation with our explanation of GREEN's data is clearly indicated.

	Acid		
	Acetic	Valeric	Caprylic
Solubility in water.	high	medium	low
Solubility in lipids	low	medium	high

When we had found the remarkable influence of fatty acid anions on the oleate coacervate (BOOIJ and BUNGENBERG DE JONG 1949) one of us (BOOIJ 1949) drew attention to the fact that the first part of the resulting curve

(Fig. 77) resembles the curves picturing the germicidal action of fatty acids
and other detergents very much. The hypothesis was given that the
detergent action on living systems is caused by the disturbance of a system
of parallelly arranged carbon chains (e. g. the protoplasmic membrane or
any other interface within the cell where phosphatides play a rôle).
Another hypothesis has been given by Aʟᴇxᴀɴᴅᴇʀ and Tʀɪᴍ (1946), who
suppose that the concentration of free (active) ions or molecules will decrease as—with the longer detergents—micelle formation sets in. It was
pointed out by Booɪᴊ and ᴠᴀɴ Lᴇᴇᴜᴡᴇɴ (1953) that it is possible to distinguish
between these two hypotheses by performing the experiments at widely

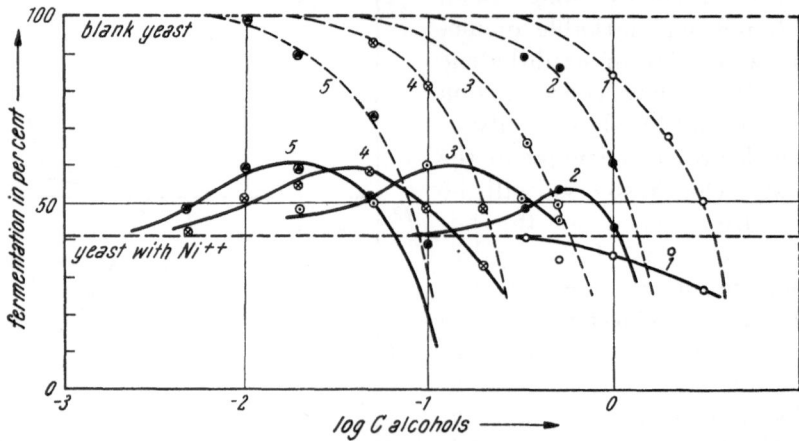

Fig. 147. Influence of the first five normal alcohols on the fermentation by yeast cells inhibited by a constant concentration of $Ni(NO_3)_2$—0.003 n—compared with the influence of the alcohols on blank yeast. The
numbers denote the chain length of the alcohols.

different detergent concentrations. In the case of Booɪᴊ's hypothesis the
maximum activity will be found at the same number of carbon atoms
(irrespective as to concentration), while according to the other hypothesis
the maximum will shift to longer chain lengths at decreasing concentrations
of the detergents. From literature it appears that both possibilities exist
in nature.

    We have seen that normal alcohols may have an opening or a condensing
influence on oleate coacervates. It has been suggested that the opening
action will come to the fore when the lipid system has a compact structure
(see p. 86). These experiments suggested some experiments on yeast cells
(Booɪᴊ and Bʀᴀɴᴅ 1953). All normal alcohols have at certain concentrations
a depressing influence on the fermentation of yeast cells. This action
follows Traube's rule. It seems probable that the alcohols penetrate into
the cell and inhibit an enzyme system there (Wᴀʀʙᴜʀɢ and Wɪᴇsᴇʟ 1912).
We asked ourselves if it would be possible to find a stimulating effect of
the alcohols on fermentation when the protoplasmic membrane has been
condensed beforehand. According to Booɪᴊ (1940) several cations (e. g.
Ni^{++}-ions) depress fermentation by displacing the normal cations of the
membrane and consequently exerting a condensing action. Indeed Booɪᴊ

and BRAND found that normal alcohols have a stimulating influence on yeast cells pretreated with $Ni(NO_3)_2$ (Fig. 147). The action of methylalcohol, which is at variance with that of the other alcohols, seems to be caused by the fact that the permeability for Ni^{++}-ions is greatly enhanced.

One of the most interesting experiments which found their provisional explanation in the complex theory concerns the influence of detergents and UO_2^{++}-ions on the fermentation of sugar by yeast cells (BOOIJ 1954). There are reasons to suppose that UO_2^{++}-ions depress the fermentation of sugar by baker's yeast by virtue of their affinity to some membrane consti-

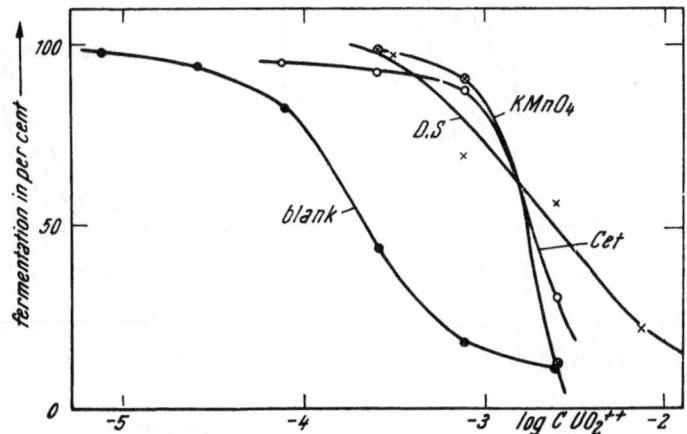

Fig. 148. The influence of $UO_2(NO_3)_2$ on the fermentation by a) blank yeast, b) yeast pretreated with do-decylsulfate (D. S.), c) yeast pretreated with $KMnO_4$ and d) yeast pretreated with cetyltrimethyl-ammonium-bromide (Cet.) The fermentation by the blank or the pretreated yeast without $UO_2(NO_3)_2$ has been given a value of 100%.

tuent (condensation of the membrane). When we treat yeast with a detergent, we see that UO_2^{++}-ions become much less toxic (Fig. 148). The detergent probably disorganize the lipid double layer, and much more UO_2^{++}-ions are then needed to condense the membrane to the same extent as the non-treated membrane. Hence the lowered toxicity. It seems practically impossible to explain these facts from the point of view that UO_2^{++} inhibits some enzyme. If one treats yeast with a low concentration of $KMnO_4$ (resulting in an unknown damage to the membrane) we get the same result: the UO_2^{++}-ions become much less toxic.

The influence of UO_2^{++}-ions on fermentation has been ascribed to a binding of one of the membrane constituents (presumably phospholipid, perhaps protein). This might have three consequences:

1. The permeability of sugar into the yeast cell is lowered (this would mean that the passive permeability of sugar into the normal yeast cell is high enough to meet the demand necessary for fermentation; a rather improbable supposition).

2. The membrane is condensed by UO_2^{++} and consequently the enzymes in the membrane responsible for the uptake of sugar cannot work properly.

3. The condensation of the membrane prevents the proper functioning of a carrier for glucose.

Here we enter the realm of active permeability. We would like to point to the interesting fact that some lipids contain glucose or galactose (which sugars are taken up quickly by many cells) and that ever more

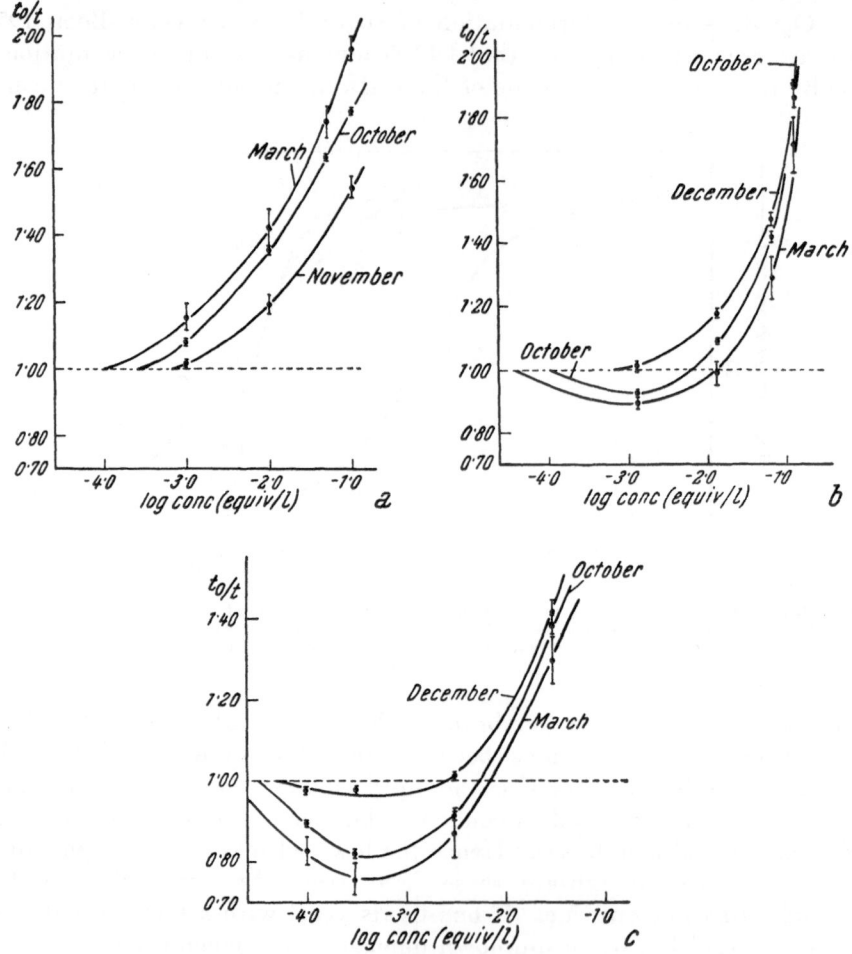

Fig. 149. The seasonal variations in the influence on the water permeability of *a* NaNO₃, *b* Ca(NO₃)₂ and *c* Co(NH₃)₆Cl₃ (t₀/t is the quotient of the time of deplasmolysis in a sucrose solution and that in the salt solution).

lipids become known which contain amino acids. It might be suggested that these lipids form an important link in active permeability. One might suggest that hexoses and amino acids are bound to membrane lipids. Then "turn over" in the double layer will occur and the substance taken up is freed by active processes. The "empty" carrier may then pick up a second molecule. The active uptake of ions too might be seen from this point of view. We still know relatively little about the lipids, but we feel sure that the study of cell lipids will reveal interesting details in future.

It would not be surprising if it were found that some of the class of fat soluble vitamins exert their activity in the protoplasmic membrane. The same might apply to some hormones. Perhaps the "active patch" theory (DAVSON and DANIELLI 1943) may also be incorporated in the complex theory.

Finally we think of the changes in permeability in ageing cells. LOEVEN (1951) investigated the influence of cations on the permeability to water of onion cells [30]. A distinct seasonal variation came to the fore (Fig. 149). This variation has been ascribed to a—hypothetical—change in the lipids of the cells. It was supposed that in winter mainly lecithin (in general electrically compensated lipids) would be present. In spring a change would take place: phosphatidic acid would be formed. It is important to note that the enzyme catalyzing this reaction (lecithinase C) has been found in plant material (HANAHAN and CHAIKOFF 1947). To test the hypothesis that the transition from a rest phase to an active phase of a living organism is accompanied by the formation of phosphatidic acid from a part of the phosphatides, MEYER [31] studied the change in lipid content in germinating soybeans. His analyses indicate that the supposed reaction does indeed take place.

We admit that none of these experiments contains on itself a direct argument for or against the complex theory. The combination of all facts, however, constitutes a very strong argument in favour of this theory. In general one might say that every other theory tries to explain only a part of the known facts. Moreover the theory has one feature which should be present in every scientific theory: it stimulates the experimental research.

References

ALEXANDER, A. E., and A. R. TRIM, 1946: The biological activity of phenolic compounds. The effect of surface active substances upon the penetration of hexylresorcinol into *Ascaris lumbricoides* var. suis. Proc. Roy. Soc. Lond. B 133, 220—234.

BOOIJ, H. L., 1940: The protoplasmic membrane regarded as a complex system. Rec. trav. bot. néerl. 37, 1—77.
— 1949: Some typical actions of fatty acids and fatty anions on biological and colloid systems. Commun. 1st Intern. Congress of Biochem. Cambridge, 253—256.
— 1949: The protoplasmic membrane regarded as a lipoprotein complex. Disc. Faraday Soc. 6, 143—152.
— 1954: Colloid-chemical contributions to the problem of biological permeability. Acta Physiol. Pharmacol. Neerl. 3, 536—552.
— and M. BRAND, 1953: The influence of alcohols on the fermentation of glucose by yeast cells. Acta Physiol. Pharmacol. Neerl. 3, 100—112.
— and H. G. BUNGENBERG DE JONG, 1949: Researches on plant growth regulators XV. Biochim. Biophys. Acta 3, 242—249.
— and A. M. VAN LEEUWEN, 1953: Influence of organic compounds on soap and phosphatide coacervates XVIII. Proc. Kon. Ned. Akad. Wetensch. Amst. B 56, 255—267.
BUNGENBERG DE JONG, H. G., A. DE BAKKER, and D. ANDRIESSE, 1955: Contributions to the colloid chemistry of phosphatides I and II. Proc. Kon. Ned. Akad. Wetensch. Amst. B 58, 239—265.

---

[30] See also p. 41.

[31] Not yet published.

Bungenberg de Jong, H. G., and J. Bonner, 1935: Phosphatide autocomplex coacervates as ionic systems and their relation to the protoplasmic membrane. Protoplasma **24**, 198—218.

Collander, R., 1926: Über die Permeabilität von Kollodiummembranen. Soc. Sci. Fennica Comm. Biol. II, Nr. 6.

— und H. Bärlund, 1933: Permeabilitätsstudien an *Chara ceratophylla*. Acta Botan. Fennica **11**, 1—114.

Danielli, J. F., and E. N. Harvey, 1935: The tension at the surface of mackerel egg oil, with remarks on the nature of the cell surface. J. cellul. a. comp. Physiol. (Am.) **5**, 483—494.

Davson, H., and J. F. Danielli, 1943: The Permeability of Natural Membranes, Cambridge.

Gay, H., and T. F. Anderson, 1954: Serial sections for electron microscopy. Science **120**, 1071—1073.

Gorter, E., and F. Grendel, 1926: On the spreading of the different lipoids from chromocytes of different animals. Proc. Kon. Ned. Akad. Wetensch. Amst. **29**, 318—320.

Green, J. W., 1949: The relative rate of permeability of the lower saturated mono-carboxylic acids into mammalian erythrocytes. J. cellul. a. comp. Physiol. (Am.) **33**, 247—266.

Hanahan, D. J., and I. L. Chaikoff, 1947: The phosphorus-containing lipides of the carrot. J. biol. Chem. (Am.) **168**, 233—240.

Loeven, W. A., 1951: Seasonal variations in the water permeability of *Allium* epidermal cells. Proc. Kon. Ned. Akad. Wetensch. Amst. C **54**, 411—420.

Michaelis, L., and A. Fujita, 1925: Untersuchungen über elektrische Erscheinungen und Ionendurchlässigkeit von Membranen IV. Biochem. Z. **161**, 47—60.

Naegeli, C. W. von, 1881: Cited in W. Pfeffer, Pflanzenphysiologie, Leipzig.

Nirenstein, E., 1920: Über das Wesen der Vitalfärbung. Pflügers Arch. **179**, 233—337.

Osterhout, W. J. V., and W. M. Stanley, 1932: The accumulation of electrolytes V. Models showing accumulation and a steady state. J. gen. Physiol. (Am.) **15**, 667—689.

Overton, E., 1895: Über die osmotischen Eigenschaften der lebenden Pflanzen- und Tierzellen. Vjschr. naturf. Ges. Zürich **40**, 159—201.

Ruhland, W., und C. Hoffmann, 1925: Die Permeabilität von *Beggiatoa mirabilis*. Planta **1**, 1—83.

Schulman, J. H., and W. McD. Armstrong, 1947: Biological activity in relation to structure and spacing of simple ionic polar groups. In Surface Chemistry, London 1949, 273—279.

Veldstra, H., and H. L. Booij, 1949: Researches on plant growth regulators XVII. Biochim. Biophys. Acta **3**, 278—312.

Warburg, O., und R. Wiesel, 1912: Über die Wirkung von Substanzen homologer Reihen auf Lebensvorgänge. Pflügers Arch. **144**, 465—488.

10. Some Remarks on the Colloid Chemistry of Protoplasm

a) Colloid morphology

Chemically speaking, the living cell consists of a great many organic and inorganic substances, several of the former belonging to the colloids. Morphologically speaking, the living cell consists of several compartments of either microscopical or submicroscopical dimensions. In the first eight chapters of this paper no attention has been paid to morphological questions. It has been tacitly supposed that the colloid systems discussed were—at least theoretically—unlimited. This was admissible as we were only interested in the interactions and properties of the kinetic units of colloid dimensions.

In the living cell the various compartments (nucleus, mitochondria, cytoplasm, vacuoles) show important differences as regards their chemical composition. Nevertheless they seem to be in equilibrium with each other.

These considerations accentuate the importance of the study of colloid systems of microscopic dimensions. This branch of colloid chemistry has been called *colloid morphology*. It studies objects consisting of one or more colloid systems, especially as regards their three-dimensional orientation. Moreover, the morphological changes caused by factors which shift the equilibrium belong to this field. The most interesting question is: to what class of colloid systems does cytoplasm belong? We will see that it is not possible to give a definitive answer to this question.

From the survey given in chapter 2 we may deduce that cytoplasm will belong to one of the following types:

1. Homogeneous submicroscopically:
   a) Sol,
   b) Coacervate.
2. Inhomogeneous submicroscopically:
   a) Sol,
   b) Coacervate,
   c) Gel.

As many experiments point in the direction of a submicroscopical structure of cytoplasm, we may safely assume that it will belong to one of the types of the second group. It must be realised, however, that the differences between the three types of the second group are often rather small. We remind of the properties of oleate systems, obtained by adding KCl to an oleate solution. The sol just before the coacervate limit and the coacervate just beyond this limit have practically the same composition and properties. The question whether a certain system should be called a elastic-viscous liquid or a gel is sometimes very difficult and the answer may be prompted by a personal predilection.

From the point of view of the forces operating between colloids (be it in sols, coacervates or gels) experimental colloid chemistry has aquainted us with two big groups of systems which are of importance for biology (1. complex systems and 2. lipophilic systems). The microscopic study of these simple models will be of interest for biology. We propose to give some examples of colloid morphological studies which have some significance for the study of living cells.

A. Disintegration of dicomplex coacervate drops in a direct current electric field

As we have already seen (chapter 4, Fig. 31) a coacervate drop consisting of two oppositely charged macromolecular colloids, shows several phenomena in a direct current electric field. Especially the disintegration phenomena have been studied extensively. From these studies it was concluded that the two colloid components are not strongly bound in a complex coacervate. They are relatively easy to displace in an electric field.

It is a well-known fact (see e. g. HEILBRUNN 1928) that protoplasm may show two phenomena under the influence of electric current: vacuolisation and disintegration. This resemblance to our coacervate drop does not prove, of course, that protoplasm should be regarded as a coacervate. It certainly suggests that the constituents of protoplasm are held together a. o. by Coulomb-forces.

B. Colloid systems surrounded by totally closed membranes

When an emulsion consisting of droplets of a mixture of gelatin and gum arabic solutions in a solution of celloidin in a mixture of ether and amylalcohol is poured on a

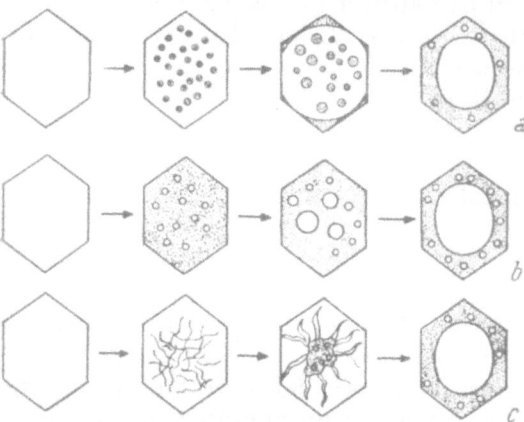

water surface, an artificial tissue of prismatic cells originates. These cells still contain the colloids and it is possible to study the influence of various factors on this "tissue" (Bungenberg de Jong, Kok, and Kreger 1940).

It is interesting to see how the same final state may be reached along several roads. In the first place one can pass along the tissue acetic acid 0.01 n (at 35⁰ C.). As soon as the pH drops below the I. E. P. of gelatin small coacervate drops begin to appear in the cell (Fig. 150 a). In the final state the complex coa-

Fig. 150. Different ways along which the final morphological position of the complex coacervate gelatin-gum arabic in the cell compartments can be reached. *a* 0.01 *n* acetic acid is passed through the cuvette at 35⁰ C. *b* 0.01 *n* acetic acid, containing 40 m. eq. p. l. NaCl is passed at 35⁰ C., followed by 0.01 *n* acetic acid without salt. *c* Cold 0.01 *n* acetic acid is passed through the cuvette and the temperature is gradually raised.

cervate wets the cell walls as a coherent liquid layer and thus encloses a central vacuole filled with equilibrium liquid. In the second experiment (Fig. 150 b) one again passes acetic through the cuvette (35⁰ C.), but this time the solution contains NaCl 40 m. eq. p. l. Nothing happens notwithstanding the fact that the pH is low enough for coacervation, as the salt suppresses the coacervation (first picture). When we now pass an acetic acid solution without salt we will observe an ever stronger vacuolisation, which finally results in the same type of system: a central vacuole surrounded by a coacervate layer. The third case may be seen in the cold. Acetic

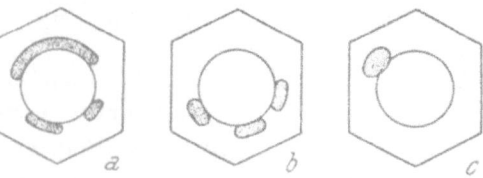

Fig. 151. Topographical position of the three coexisting liquids (after complex coacervation of a gelatin + arabinate + nucleate sol mixture) in the cell compartments of an artifical tissue (schematic).

acid 0.01 n is again passed through the cuvette. We do not see morphological changes as the gelated sol mixture gradually changes into a complex gel. When, however, the temperature is raised gradually the complex gel weakens and changes into the ordinary two layer system (Fig. 150 c). Thus the three possibilities are: 1. origination of a coacervate in a sol, 2. appearance of vacuoles in a coacervate rich in water and 3. softening of a complex gel.

If a mixture of gelatin, gum arabic and nucleate is enclosed in colloidin cells, we may obtain three coexisting liquids. A coacervate consisting mainly of gelatin and gum arabic surrounds a central vacuole (equilibrium liquid). In this coacervate a second coacervate is embedded, containing mainly gelatin and nucleate (grey drops in Fig. 151). The resemblance to a plant cell (Fig. 151 c) is striking. We need not repeat that this morphological similarity is no proof for the coacervate nature of cytoplasm of the mature cell. One might also consider the cytoplasm as a sol, but then a rather rigid membrane must be postulated at the boundary cytoplasm/ vacuole. The nucleus might be regarded as a gel or again as a sol, provided that in the last case a strong membrane separates nucleus and cytoplasm. We think it probable, however, that the nucleus at least in the first stage after cell division has a coacervate nature. Afterwards it might—but need not—change into a gel or a sol surrounded by a new membrane. The cytoplasm of the young plant cell too seems to be a coacervate. During the course of its development we observe the appearance of many small vacuoles, which finally unite into a central vacuole. This phenomenon could not occur in a sol and is practically impossible in a gel. Provided a strong membrane separates cytoplasm and vacuole, cytoplasm might be a sol in the case of the mature cell. It would not be surprising if it should be found that cytoplasm can be—according to circumstances—either a sol, a coacervate or a gel with a submicroscopical structure.

C. Accumulation of dyes

The dyeing of cells of *Allium cepa* by neutral red depends very much on pH. At $pH = 5–6$ only the cell wall will be coloured. At higher values of pH, however, only the vacuole takes up the red dye, while the rest of the cell (cytoplasm as well as cell wall) remains uncoloured. When the cell is killed (e. g. in hot water) cytoplasm and nucleus will show a strong red colour when the cell is placed in a dilute solution of neutral red.

The dyeing of the cell wall has to be regarded as a reaction between the positive dye-ions and the negative (carboxyl) groups of the cell wall. This follows from the fact that the cell wall will not be coloured either at high pH (there no dye-cations will be present) or at low pH (there the carboxyl-group will have no charge). The dyeing of the vacuole too may be understood from simple considerations. Here the permeability of the cytoplasm plays a decisive rôle. Neutral dye molecules will permeate very much faster than dye-cations. Thus a coloration of the vacuole will take place only at relatively high pH of the medium. Moreover, the vacuole sap must contain negatively charged colloid to bind the dye-cations (which will originate after permeation of the molecules).

This phenomenon may be easily demonstrated when using gelatinised hollow spheres (BUNGENBERG DE JONG, BANK, and HOSKAM 1940). When negative coacervate drops (gelatin and arabinate) are brought into contact

with a large amount of distilled water, a hollow sphere will appear via
stages of strongly vacuolized spheres (Fig. 152). On cooling these hollow
spheres gelatinize. When they are then placed in a very dilute solution of
neutral red (after a preliminary treatment with tap water), an accumu-

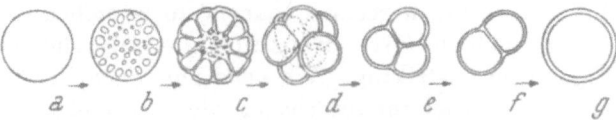

Fig. 152. Morphological development of the hollow spheres from negatively charged coacervate drop of the
complex coacervate gelatin/gum arabic.

lation of neutral red in the vacuole takes place, while the surrounding shell
itself remains practically uncoloured (Fig. 153). After the accumulation
has reached a certain degree the very intensely coloured vacuole liquid
"demixes." A number of very small intensely coloured drops separate out
and finally the shell becomes covered with a thin layer of coloured liquid.

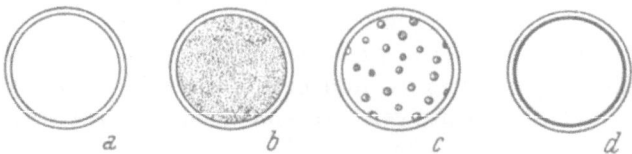

Fig. 153. Accumulation of a basic dye and coacervation in the central vacuole of a gelatinised hollow sphere.
a Original state. b Diffuse accumulation of the basic dye. c Coacervation of the arabinate by the dye cations. d The intensely coloured coacervate concentrates along the inner wall of the hollow sphere.

The preceding treatment with tap water—increase of the pH—is necessary condition for the success of the experiment. The complex relations
in the complex gel are lost and the vacuole liquid becomes a sol of gum
arabic. With neutral red a new, strongly coloured, complex coacervate
(arabinate + dye cation) is formed within the vacuole.

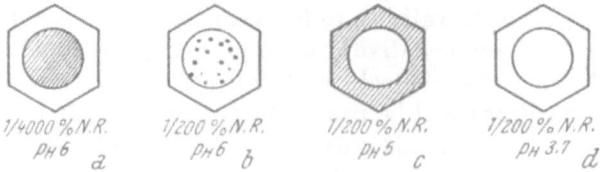

Fig. 154. Reversible colloid displacement between gelatinised complex coacervate on the wall and central
vacuole (see text).

The fact that cytoplasm of the living cell, in contradistinction to that
of the killed cell, is not coloured by neutral red is a problem of greater
importance. Here too model studies may throw some light on the question.
Bungenberg de Jong and Bakhuizen van den Brink (1947) studied the
accumulation of neutral red in a coacervate enclosed in walls of celloidin
(Fig. 154).

If we make a coacervate of gelatin and gum arabic (see Fig. 150) and pass a cold dilute buffer of pH $= 3.7$ to which neutral red has been added ($^1/_{200}$%), neither the gelatinized coacervate nor the vacuole take up the colour (Fig. 154 d). At pH $= 6$ and a low concentration of neutral red ($^1/_{4000}$%) a diffuse accumulation will be seen in the vacuole (Fig. 154 a). At the same pH, but a higher neutral red concentration, dark coloured drops will be formed in it (stage b). Starting from stage b and passing dilute buffer of pH $= 5$ with $^1/_{200}$% neutral red, we will observe a disappearance of the intensely coloured drops, while the gelatinized coacervate becomes red (Fig. 154 c).

For an explanation of these phenomena it is important that: 1. gum arabic when present alone in the cell compartments, accumulates neutral red at all pH's mentioned here (even at pH $= 3.7$) and 2. the I. E. P. of the gelatin used is approximately 5.

The failure of the staining at pH $= 3.7$ (Fig. 154 d) must be due to the fact that all negative charges of the gum arabic are taken up by the positive charge of gelatin. In this case the gum arabic will not be found in the vacuole. At pH $= 6$ the gum arabic will gradually diffuse into the vacuole. Thus a gradual accumulation of neutral red in the vacuole will take place at this pH (Fig. 154 a and b). Lowering the pH from 6 to 5 evidently makes all gum arabic disappear from the vacuole. Thus it must be supposed that the gum arabic is more or less bound to the gelatin, though the latter is at its isoelectric point. The remaining negative groups of the gum arabic can then bind neutral red cations.

In living protoplasm the situation may be compared with Fig. 154 a or b. We must conclude that there too no negative groups are free to combine with the dye-cations. This causes some surprise as—at the pH prevailing in cytoplasm—most colloids will bear a negative charge. We come to the important conclusion that most of the negative groups present in cytoplasm are not freely accessible. Presumably they take part in complex relations. Moreover, a certain amount of salt will inhibit the binding of cationic dyes to the excess of negative groups. In the living cell the colloids form a submicroscopical structure. If this structure is destroyed, the salt will diffuse from the cell, the negative groups will become free and the killed cytoplasm can be strongly dyed.

These three examples may serve to understand the aim of colloid morphology (see also the chapter "Morphology of Coacervates" by BUNGENBERG DE JONG in KRUYT's Colloid Science II, 1949).

b) The submicroscopical structure of cytoplasm

As this handbook will contain many pages dealing with the properties of cytoplasm, we will remind only of those characteristics which have a direct bearing on our subject. In many cases protoplasm seems to be a liquid. It is customary to refer in these cases to the "sol properties" of protoplasm. We have already stressed that coacervates are liquid too. Thus one might equally well speak of "coacervate properties." In the following pages we will join in the customary nomenclature, but it should be remembered that the flow of protoplasm, its large water content, the relatively low viscosity, the convex shape of the plasmolysed cell, etc. constitute no proof of its sol character. We can only subscribe to the view that in these cases the protoplasm is not a gel in the ordinary sense.

In contradistinction to water, glycerol or other normal liquids, protoplasm shows an anomaly as regards its viscosity. The viscosity depends on the shearing stress; the higher the shearing stress, the lower the viscosity. This fact points conclusively in the direction of a submicroscopic structure. As we have seen (p. 65) it does not prove that protoplasm is a gel; the strongest argument in favour of this view would be the demonstration of a yield value.

Often protoplasm does not have the properties of a liquid of low viscosity. It may show plastic properties and even demonstrable elasticity. Plasmolysis often is not convex, but concave or angular, which indicates a certain rigidity of the cytoplasm. Moreover cytoplasm exhibits a spinning capacity. Long strands can be drawn from it. None of these phenomena, however, give unquestionable arguments for the view that here one must speak of the gel state of protoplasm. Certainly protoplasm may vary a great deal as regards its rigidity. The very important, reversible changes in rigidity are commonly known as the sol-gel transition, though it may be asked whether the more rigid state is really a gel or if the less rigid state is really a sol. However, the remarkable properties of protoplasm cannot be understood without the assumption of a submicroscopic structure; it is neither an ordinary sol, nor an ordinary coacervate, nor an ordinary gel.

In an endeavour to explain the properties of protoplasm Frey-Wyssling developed his theory of junctions. In this theory the proteins are considered to be the structural elements of the cytoplasm. The side chains of the proteins show several possibilities of attraction, which have been classified in four groups:

I. Homopolar cohesive bonds (London-Van der Waals forces) ; e. g. between the CH_3-groups of alanine and leucine.

II. Heteropolar cohesive bonds (dipole/dipole attraction and hydrogen bonds); e. g. between the hydroxy-groups of serine.

III. Heteropolar valency bonds (Coulomb forces) ; e. g. between oppositely charged side chains of the proteins.

IV. Homopolar valency bonds; for instance S-S-bridges between cystein side chains.

Thus the protein macromolecules in protoplasm are supposed to be interlinked to form a framework. Various agents may disrupt these junctions. The main point of this theory is the idea that the protein molecules may be released reversibly; the junctions do not have a static, but a dynamic character. In this way the junction theory tries to explain that protoplasm is a co-ordinated whole, while on the other hand it shows some characteristics of liquids.

The criticism of the theory of junctions should be divided into two groups of objections. First of all there are some authors who object to the idea of the junctions as such. Höfler e. g. found that in cap-plasmolysis of *Allium* cells the protoplasm may swell up to 10 or more times its original volume. He concluded that no framework could be present, as the structural elements would be pushed so far apart, that they would completely change their mutual relations. This criticism may be rejected on the ground of the fact that every gel-like colloid system may show the property of limited swelling. Moreover, cap-plasmolysis is easily reversible. It points, however, to the second class of. criticism. The swelling of a protein gel to many times its original volume is only possible when the junctions are fairly strong. The amount of water between the meshes of the protein framework is so large that only relatively few (and consequently strong) junctions exist. The junctions proposed by Frey-Wyssling do not—with the exception of the last group—have this quality. Especially the first group of attraction (London-Van der Waals forces) is very weak, when operating between small hydrocarbon groups. This means that the time of contact between these groups will be extremely low. Now the difficulty lies in the fact that a much higher duration of contact is necessary in order to observe a visible elasticity. Even the third group (Coulomb forces at isolated spots) does not have this quality. This can be demonstrated by the fact that a gelatin-gum arabic coacervate (at 40⁰ C.) does not show the slightest trace of elasticity, notwithstanding the fact that the number of "junctions" is high enough to retract the colloids from the equilibrium liquid. The coacervate behaves like a normal liquid.

In order to explain the extraordinary properties one should look for junctions of a medium duration time; the time should be long enough to ensure the visible elasticity and short enough to give protoplasm its fluid character. In a gelatin gel e. g. the time of duration of the junction is too long; it is a very elastic but solid system. As regards the theory of junctions of protoplasm as given by Frey-Wyssling we may say that the first three classes of attractions have a too short time of duration, while in the last class the time of duration is too long. If one likes to regard the proteins as the essential structural elements two possibilities still remain.

1. One might postulate the existence of special spots on the structural proteins. These spots would have to consist of a number of side chains, able to give attractions of the first three types. Thus a summation of a number of weak forces would result in a junction of medium duration. These specific groupings of amino acids do not seem impossible, but rather improbable.

2. On the other hand one might suppose that only the strongest bonds constitute the junctions between the structural proteins. Then the S-S-bonds are the sole junctions of importance. This would mean of course, that the sol-gel transformation would depend solely on local changes in the redox-potential. In view of the influence of cations on the rigidity of protoplasm this seems an unlikely idea.

Whereas no artificial protein system has been described which even remotely shows the peculiar properties of protoplasm, elastic-viscous systems of association colloids became known, which have some highly interesting features. We dealt with these systems in chapter 6 and we will remind of the following properties: 1. the viscosity depends on the shearing stress, 2. the systems show clearly visible elasticity, 3. birefringence of flow can be observed, 4. they show a spinning capacity, 5. a yield value cannot be demonstrated, and 6. they are—in contrast to gelatin gels— quickly and completely reversible. This resemblance led to the supposition that the framework of protoplasm is a system of micelles of association colloids. According to this hypothesis the natural association colloids, the phosphatides, are of the utmost importance for the submicroscopic texture of protoplasm. It should be noted that the criticism against the protein framework (viz. the Coulomb forces at isolated spots are not strong enough to explain a visible elasticity; p. 55) does not apply in this case, as we are dealing here with Coulomb interactions between compact charge mosaics on the surfaces of the sandwich micelles.

Strong circumstantial evidence for this view may be deduced from the well known experiments on the influence of cations on the rigidity of cytoplasm. Cells of *Allium cepa* are placed for 24 hours in a weak KCl solution or a weak $CaCl_2$ solution respectively. Then the cells are plasmolyzed with the aid of a hypertonic sucrose solution. Then the former cells will show convex plasmolysis, the latter on the other hand concave plasmolysis. Thus in the latter case the rigidity of the cytoplasm (at least its periphery) is much higher.

This difference between calcium- and potassium-ions is also clearly indicated in experiments on cap-plasmolysis. If a plasmolysis brought about by KNO_3 (or another alkali-salt) is continued for a long time the protoplasm will swell up considerably. This phenomenon can not be understood from considerations on osmosis, as the swollen protoplasm will gradually return to its original volume, when the KNO_3 solution is replaced by an isotonic $CaCl_2$ solution.

The phenomena described can be explained from the hypothesis that normal protoplasm contains calcium-ions. These cations are of fundamental importance, but they can be replaced by alkali-ions. Then the coherence of the cytoplasm is weakened (the number or the strength of the junctions decrease) and we will observe a swelling of the cytoplasm. According to this view this phenomenon should be reversible.

We may now ask ourselves which charged group in the biocolloids may be responsible for the large difference between calcium- and potassium-ions. We then look back to chapter 3, where we have seen that the phosphate group is the only negative group which shows a large difference in affinity as regards these two ions. Consequently we may say that biocolloids with phosphate groups must play an important rôle in the structure of cytoplasm. It seems self-evident to ascribe this rôle to the phosphatides.

Thus we come to the conclusion that the hypothesis that proteins are the structural elements of cytoplasm has only a weak foundation. It is suggested that phosphatide micelles play the rôle of the structural elements in cytoplasm. It must be admitted, however, that much work has still to be done to give this hypothesis a sound basis. In the course of this study it might also be found that both proteins and phosphatides are essential for the submicroscopic structure. In view of the interactions between linear proteins and ionized lipids (chapter 8) this would not be very surprising.

c) Protoplasmic flow

It is self-evident that the ideas on protoplasmic flow are strongly influenced by the opinions formed on the submicroscopic structure of cytoplasm. In cells with amoeboid movement protoplasmic flow is accompanied by continuous sol-gel transitions. In plant cells the whole protoplasm rotates along the cell wall (cyclosis). In some cases cytoplasmic strands can be seen moving across the central vacuole. If one takes the point of view of a protein framework, the only possibility to account for cyclosis seems to be the postulation of contractibility in this framework (FREY-WYSSLING). A submicroscopic part of the protoplasm

Fig. 155. Streaming in a coacervate drop and direction of creep in a diffusion field.

gelates and contracts for a short time. Then relaxation follows and the adjoining region contracts. Thus a wave of contraction moves periodically along the protoplasmic strand. It is difficult to believe in this hypothesis. The objections against the theory of junctions between proteins carry even more weight in this case. Moreover, a mobility caused by contraction

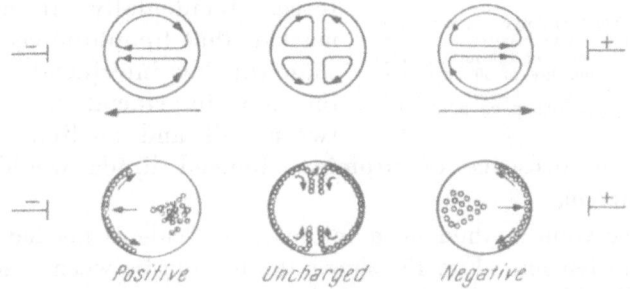

Fig. 156. Streaming and vacuolation phenomena in a complex coacervate in a d. c. electric field. Upper row: streaming phenomena and direction of creep movements. Lower row: vacuolation a short time after applying the field.

needs a solid substratum. Admitting this, it is surprising that flow may proceed simultaneously in opposite directions.

We will not enter into a discussion of the extensive literature on this problem, but relate some experiments on microscopic colloid systems, which give rise to stimulating ideas. A coacervate drop is brought on a microscope slide, after which a diffusion field is produced in the equilibrium liquid (BUNGENBERG DE JONG and HOSKAM 1941). Then one can observe the phenomena pictured in Fig. 155. When a substance, which makes the coacervate richer in colloids ("condensing" action) diffuses from the right to the left, we see: 1. vacuolisation at the right end of the drop, 2. flow of the superficial coacervate layers to the right (as seen from the motion of the vacuoles carried along) and, as a result of this, 3. "creeping" of the coacervate on the glass surface.

In an electric (direct current) field similar observations can be made (Fig. 156). The flow phenomena can be easily observed already in weak fields (5–10 Volt/cm.). In the original publication it has been assumed that the pH left and right in the drops gets a different value as a consequence of the field. In a later investigation (De Ruiter and Bungenberg de Jong 1947) it was shown that the ratio gelatin/gum arabic changes locally. This results in a local change in interfacial tension.

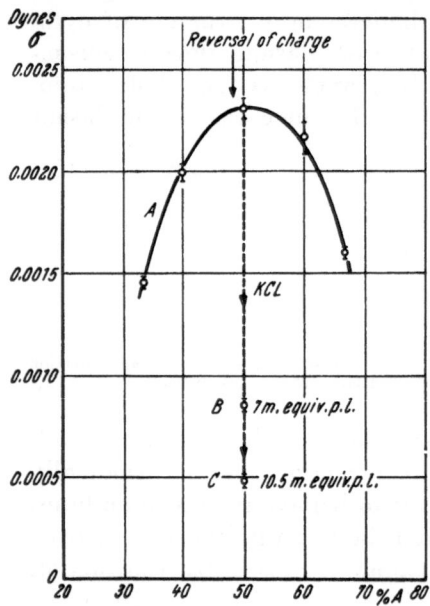

Fig. 157. The interfacial tension coacervate/equilibrium liquid in the complex coacervate gelatin/gum arabic. A Influence of the mixing ratio (e.g. 60% A means a mixture of 60 parts gum arabic 2% and 40 parts gelatin 2%). B and C Influence of KCl (7 and 10.5 m. eq. p. l.) on σ.

At first sight this seemed surprising as the interfacial tension coacervate/equilibrium liquid is low (as judged from the experiments in bulk, where the interface is always absolutely flat). It is not possible to measure this interfacial tension with a Du Noüy tensiometer as the platinum ring is pulled out of the interface with the slightest force. Incidentally, it may be remarked that lipophilic coacervates too show this low interfacial tension. Thus the low interfacial tension e. g. between cell and medium need no be ascribed to the presence of proteins. Ionized lipids would show the same phenomenon.

A capillary tube method on a microscopic scale is needed to measure this interfacial tension. Fig. 157 shows the tension between coacervate and equilibrium liquid as a function of the mixing ratio of gelatin and gum arabic (De Ruiter and Bungenberg de Jong 1947). Extraordinarily small values have been obtained with the aid of this method. Obviously, the interfacial tension is maximal at the reversal of charge point. Moreover, every factor which intensifies the complex relations causes the interfacial tension to increase. Conversely, the factors which decrease the complex relations make the interfacial tension decrease (see e. g. the influence of small amounts of KCl in Fig. 157).

The pictured flow phenomena (Figs. 155 and 156) can be easily understood from these findings. We will find a flow along the surface of coacervate drops from a region of low interfacial tension to a place with a higher interfacial tension. Then, of course, we will find a compensating flow in the middle of the drop.

These phenomena suggest that in most cases protoplasmic flow will be the result of local variations in surface tension. Extremely small differences in interfacial tension will, as has been shown, be sufficient for a flow which can be observed microscopically. Streaming in opposite directions may

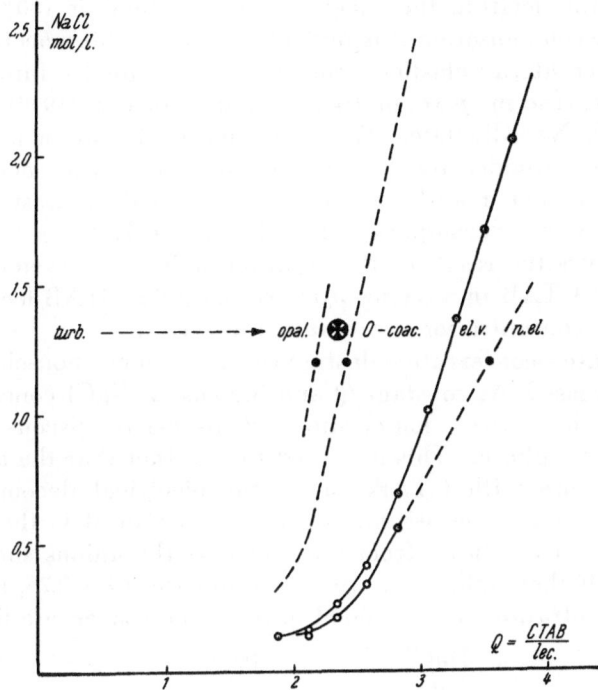

Fig. 158. Colloid systems obtained in the combination lecithin, cetyltrimethylammoniumbromide (CTAB) and NaCl.

come to the fore, as in some cases a compensating flow is necessary. The driving force of the protoplasmic flow can be of different origin, e. g. a diffusion field in an intact tissue, local variations of metabolism (resulting in different productions of a swelling or condensing substance) or local differences in membrane potential.

In this respect the hypothesis that the structural elements of protoplasm consist of phosphatide micelles tallies well with the idea that small variations in interfacial tension cause protoplasmic flow. We remind of Fig. 142 K, where we have seen that in an electric d. c. field (already at low intensity) streaming in a phosphatide coacervate may be observed. As we have seen in chapter 6 the lipophilic elastic-viscous systems do not show a yield value. In other words, *the slightest shearing stress already causes flow*. From a theoretical point of view this is of great importance, as the absolute values of interfacial tension are extraordinarily low. At first sight one would be inclined to regard the smallness of these values as an objection against our hypothesis. This objection may be overruled, however, in systems showing no yield value. It would be worth while to examine whether protoplasm shows a yield value or not.

Finally we will describe some experiments by Bungenberg de Jong and De Bakker, which perhaps might give some hints as regards the problem of the reversible sol/gel transitions in protoplasm. It has already been discussed, that lecithin, when brought in water, forms a suspension of smetic particles. In this lecithin the "degree of occupation" is 100% (see p. 52). By electrical decompensation it is possible to change this smectic suspension into an O-coacervate, an elastic-viscous system and finally into a non-elastic system. Bungenberg de Jong, de Bakker, and Andriesse (1955) succeeded in doing this with Na-salicylate. We have suggested that in nature the decompensation is provided by the presence of phosphatidic acid. As phosphatidic acid is not readily available, it seemed interesting to study phosphatide systems decompensated with laurylsulfate or CTAB.

Fig. 158 shows the result of the addition of NaCl to systems consisting of lecithin and CTAB in varying proportions ($Q = \text{CTAB/lecithin}$). At a constant NaCl concentration we obtain the series:

smectic phase—coacervate—elastic-viscous system—non-elastic system, when Q is increased. At constant Q and increasing NaCl concentration we pass through the series: non-elastic system—elastic-viscous system—coacervate—smectic phase. This is caused by the fact that the added anions combine still more with CTAB. Thus the electrical decompensation is gradually abolished. The assumption that this indeed is the background of the salt influence follows from the fact that the anions show a typical series as regards their influence. In a certain case ($Q = 2.75$, at 19° C.) the following concentrations were needed to reach the coacervate limit:

$$
\begin{array}{lll}
\text{NaCl} & . \quad . \quad . \quad . & 0.6\,n, \\
\text{KBr} & . \quad . \quad . \quad . & 0.03\,n, \\
\text{NaNO}_3 & . \quad . \quad . & 0.018\,n, \\
\text{KJ} & . \quad . \quad . \quad . & 0.005\,n, \\
\text{KCNS} & . \quad . \quad . & 0.002^5\,n.
\end{array}
$$

This anion series is typical for the affinity of anions to the positive groups found in CTAB:

$$\text{CNS} \rangle \text{J} \rangle \text{NO}_3 \rangle \text{Br} \rangle \text{Cl}.$$

This finding is of great importance for biology. It suggests that in the natural mixture of lecithin and phosphatidic acid we will find a cation series which represents the affinity for the phosphatidic acid.

In a direct current electric field these coacervates show remarkable properties (most pronounced at the left half of the coacervate region). In a field of 3 Volt/cm. nothing particular happens. A slight vacuolisation occurs and a flow appears exactly as expected in the case of a positive drop. The vacuoles behave as they do in every ordinary coacervate.

In a stronger electric field we see the appearance of a thickening of the wall (Fig. 159 *a, b, c*) at the side of the positive electrode. This thick and rather rigid wall shows double refraction with the long axis of the index ellipse perpendicular to the surface. The radius of this part of the drop is smaller than that of the right part.

If the electric field is switched off, the thickening disappears in a few minutes. The double refraction too will fade away. Thus the phenomenon seems to be completely reversible.

If the field is maintained, myelin tubes will suddenly grow out of the surface (Fig. 159 *d*). Here too the long axis of the index ellipse stands perpendicular to the surface. A short myelin tube will be retracted, when the field is switched off. Under the same conditions a long myelin tube

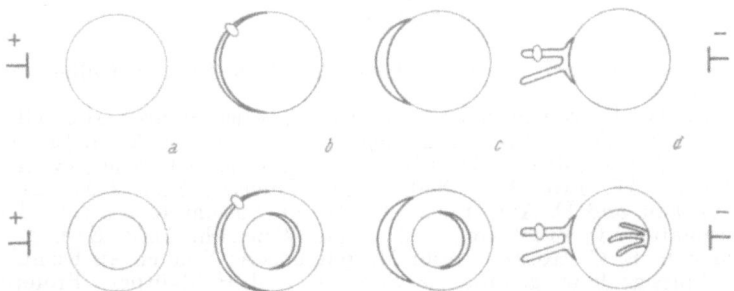

Fig. 159.   Thickening of the wall of a lecithin/CTAB coacervate and outgrowth of myelin tubes under the influence of a d. c. electric field.

loses its rigid character; if it by chance touches the surface of the coacervate drop, it will fuse with the drop and gradually disappear.

When a vacuole is present, we see in an electric field the same phenomena at the border coacervate/vacuole (Fig. 159). First a thickening of the wall occurs, followed by an outgrowth of myelins.

The explanation of these phenomena must be sought for in the influence of the electric field on the ratio CTAB/lecithin (Q). In such a field Q will decrease at the side of the drop turned to the positive electrode; Q will increase at the other side. The surface tension at the left increases, while that at the right decreases. The usual flow within the coacervate drop starts. When, however, the quotient Q is decreased sufficiently, the border between coacervate and smectic phase (Fig. 158) will be crossed. At the left of the drop this new phase separates out (Fig. 159). If the field is switched off the original ratio between CTAB and lecithin is restored throughout the drop.

We might describe the smectic phase—coacervate transition as a reversible decompensation of the lecithin system. The botanist will think of the outgrowth of myelin tubes in the vacuole of plant cells, which may sometimes be observed at beginning plasmolysis. Here we might think of a local disruption of the protoplasmic membrane, resulting in a locally increased intrability or in a local electric field. Then a shift of the decompensating substance in the phosphatide system might be visualized.

On the basis of the experiments quoted we are inclined to regard the

162   I, 2: H. L. Booij and H. G. Bungenberg de Jong, Biocolloids

sol-gel transition of protoplasm as a reversible change in the electric compensation of the phosphatide micelles presumably forming the structural elements of protoplasm.   One of the many ways in which this reversible compensation might be effected would be the presence of a local electric field.   Another way would be the production of a substance chelating calcium ions.   Then this substance would withdraw these ions from the phosphatide micelle, which would result in an electrical decompensation. Destruction or inactivation of this hypothetical substance would lead to compensation.   A more direct method would be a biochemical process by which phosphatide would be reversibly converted into phosphatidic acid. Here lies an enormous field of experimental possibilities.

References

Bungenberg de Jong, H. G., 1949: Chapter XI in Kruyt's Colloid Science II. Amsterdam.
— and R. C. Bakhuizen van den Brink, 1947: Tissues of prismatic celloidin cells containing biocolloids. Proc. Kon. Ned. Akad. Wetensch. Amst. 50, 436—441.
— and A. de Bakker, 1956: Contributions to the colloid chemistry of phosphatides III and IV. Proc. Kon. Ned. Akad. Wetensch. Amst. B 59, 124—161.
— A. de Bakker, and D. Andriesse, 1955: Contributions to the colloid chemistry of phosphatides I. Proc. Kon. Ned. Akad. Wetensch. Amst. B 58, 238—250.
— O. Bank und E. G. Hoskam, 1940: Morphologische Studien an Komplexkoacervaten. Flüssige bzw. gelatinierte Schaum- und Hohlkörper. Protoplasma 34, 30—54.
— and E. G. Hoskam, 1942: Motory phenomena in coacervate drops in a diffusion field and in the electric field. Proc. Kon. Ned. Akad. Wetensch. Amst. 44, 1099—1103.
— B. Kok, and D. R. Kreger, 1940: Tissues of prismatic cells containing biocolloids I. Proc. Kon. Ned. Akad. Wetensch. Amst. 43, 512—521.
Frey-Wyssling, A., 1953: Submicroscopic morphology of protoplasm. Amsterdam.
Heilbrunn, L. V., 1928: The colloid chemistry of protoplasm. Berlin.
Ruiter, L. de, and H. G. Bungenberg de Jong, 1947: The interfacial tension of gum arabic-gelatine complexcoacervates and their equilibrium liquids. Proc. Kon. Ned. Akad. Wetensch. Amst. 50, 836—848.
— — 1947: Contribution to the explanation of motory and disintegration phenomena in complex coacervate drops in the electric field. Proc. Kon. Ned. Akad. Wetensch. Amst. 50, 1189—1200.

Protoplasmatologia. Handbuch der Protoplasmaforschung.

Herausgegeben von **L. V. Heilbrunn**, Philadelphia, und **F. Weber**, Graz

Das Handbuch erscheint in selbständigen Einzelveröffentlichungen, die zu 14 Bänden vereinigt werden. Jeder selbständig erscheinende Handbuchteil ist einzeln käuflich. Bei Verpflichtung zur Abnahme des gesamten Handbuches, bei Vorbestellung der einzelnen Teile sowie für Abonnenten der Zeitschrift „Protoplasma" ermäßigt sich der Preis um 20%. Über die Disposition des Gesamtwerkes und die nächsten Veröffentlichungen gibt der Verlag bereitwilligst Auskunft.

Zuletzt erschienen:

Die submikroskopische Struktur des Cytoplasmas. Von Prof. Dr. A. Frey-Wyssling, Institut für Allgemeine Botanik der Eidg. Technischen Hochschule Zürich. Band II. Cytoplasma. A. Morphologie. 2. Mit 90 Textabbildungen. IV, 244 Seiten. Gr.-8°. 1955. S 255.—, DM 42.50, sfr. 43.50, $ 10.10

The pH of Plant Cells. By Prof. Dr. James Small, The Queen's University Belfast, Department of Botany. With 3 figures. IV, 116 pages. — **The pH of Animal Cells.** By Professor Dr. Floyd J. Wiercinski, Hahnemann Medical College, Department of Physiology, Philadelphia, Pa. With 7 figures. 56 pages. **Band II. Cytoplasma.** B. Chemie. 2. Spezielle Cytochemie und Histochemie. c. Gr.-8°. 1955. S 270.—, DM 45.—, sfr. 46.—, $ 10.70

The Metachromatic Reaction. By Prof. Dr. John W. Kelly, Department of Anatomy, Medical College of Virginia, Richmond, Virginia. **Band II. Cytoplasma.** D. Vitalfärbung, Vitalfluorochromierung. 2. With 5 figures. IV, 98 pages. Gr.-8°. 1956. S 210.—, DM 35.—, sfr. 35.90, $ 8.35

The Enzymology of the Cell Surface. By Aser Rothstein, Rochester, New York. With 21 figures. IV. 86 pages. — **Tension at the Cell Surface.** By E. Newton Harvey, Princeton, New Jersey. With 13 figures. 30 pages. **Band II. Cytoplasma.** E. Cytoplasma-Oberfläche. 4. 5. Gr.-8°. 1954. S 168.—, DM 28.—, sfr. 28.80, $ 6.70

Chemistry and Physiology of Mitochondria and Microsomes. By Olov Lindberg, Ph. D., and Lars Ernster, Ph. D., beide Wenner-Gren's Institute, Stockholm. **Band III. Cytoplasma-Organellen.** A. Chondriosomen, Mikrosomen, Sphaerosomen. 4. With 32 figures. IV, 136 pages. Gr.-8°. 1954. S 204.—, DM 34.—, sfr. 34.80, $ 8.10

Le vacuome de la cellule végétale. Morphologie. Par Prof. Dr. Pierre Dangeard, Laboratoire de Botanique, Faculté des Sciences, Université de Bordeaux. Avec 26 figures. IV, 41 pages. — **Le vacuome animal.** Par Raymond Hovasse, Professeur à la Faculté des Sciences de Clermont-Ferrand, France. Avec 16 figures. 37 pages. — **Contractile Vacuoles of Protozoa.** By Prof. Dr. J. A. Kitching, Department of Zoology, University of Bristol. With 20 figures. 45 pages. — **Food Vacuoles.** By Prof. Dr. J. A. Kitching, Department of Zoology, University of Bristol. With 24 figures. 54 pages. **Band III. Cytoplasma-Organellen.** D. Vacuome. 1. 2. 3 a, 3 b. Gr.-8°. 1956. S 402.—, DM 67.—, sfr. 68.80, $ 16.—

Active Transport through Animal Cell Membranes. By Dr. Paul G. LeFevre, Medical Branch, Division of Biology and Medicine, United States Atomic Energy Commission, Washington, D. C. **Band VIII.** Physiologie des Protoplasmas. 7. Aktiver Stofftransport. a. With 31 figures. IV, 123 pages. Gr.-8°. 1955. S 228.—, DM 38.—, sfr. 38.70, $ 9.—

Red Cell Structure and Its Breakdown. By Prof. Dr. Eric Ponder, The Nassau Hospital, Mineola, N. Y. **Band X. Pathologie des Protoplasmas.** 2. With 58 figures. IV, 123 pages. Gr.-8°. 1955. S 240.—, DM 40.—, sfr. 40.90, $ 9.50

Protoplasmatische Pflanzenanatomie. Von Dr. Lotte Reuter, Privatdozent an der Universität Wien. **Band XI. Vergleichende Protoplasmatik.** 2. Mit 64 Textabbildungen. IV, 131 Seiten. Gr.-8°. 1955. S 204.—, DM 34.—, sfr. 34.80, $ 8.10